Brigadier General Lecel Lee
The Toughest Air Force Commander in World War II

☆☆☆

Lecel Lee from Parker County, Texas, made gallant and heroic contributions in bringing the war to an end. He was, one of a kind when it came to combat missions. Unlike most Wing Commanders, Lee actually flew missions and led formations into combat. The 49th Bomb Wing strategic bombing mission during the "final knock-out" blow to Ploesti, Rumania was led by General Lecel Lee flying as Command Pilot in the lead B-24 Liberator aircraft.

Major General Twining, 15th Air Force Commander, "It gives me much pleasure to pass on this tribute. I realize that our successes at Ploesti would not have been possible except for the skill and industry with which all of you, in both air and ground echelons, performed your assigned tasks. By your work you have made major contributions to an early defeat of the German war machine."

Parker County, Texas

History, and enumeration of the ups and downs in the life of a local World War II hero

☆

This book reflects the life of General Lecel Lee from his birth on the East side of Weatherford, Texas, July 18, 1903, until his death, February 26, 1976. Many pictures are included of Parker County where his father ran one of the first dairies in the area. The World War II history is portrayed through the oral history and diaries of Brigadier General Lecel Lee, as well as many Newspaper Articles and information from "All Over"

This is the fifth book in a series of books by Leon Tanner and Mary Kemp, which include *All True History*, spotlighting Weatherford and Parker County, Texas.

Nebo Valley Press
Leon Tanner- Mary Kemp
2603 S. Main St
Weatherford, Texas, 76087

Lois Johnson, Writer/Editor

Leon Tanner
Mary Kemp

Proceeds from this book will be shared with Parker County Charities.

ISBN-0-9726133-3-1

The Book Makers

Leon Tanner started all of this World War II story because Leon and three of his brothers, Noah, Ed and Harold, not only served in World War II, but were neighbors of the Lee Family while living on the east side of Weatherford for years. In addition to being friends for years, another brother -Tom Tanner- married Lecel's sister, Emily. Emily had a daughter, Mary India Moorehead, who married Leroy Schoonover and they have three children, Emily, Becky and Ben.

Now Leon contacted Mary India, born and raised in Parker County, and found she had picture books galore along with newspaper write ups about Lecel. The book was started and we visited the Eisenhower Library in Kansas to get copies of Lecel's Diaries from 1937-1950.

The diaries contain nearly 1500 pages. Documents from Dwight D. Eisenhower Library Oral History interviews with General Lecel Lee, Alan Thompson, Maclyn Burg and William Jones were also acquired at that time. We would have to say that Leon Tanner, and Mary India Schoonover, are the original "Book Makers."

Leon Tanner and Mary Kemp of Nebo Valley Press started the book. But, due to health reasons and work on other books, the writing progress was stalled. Tanner and Kemp hired others to help, but that did not work out.

Once again we turned to our neighbor, Lois Johnson for help. Because of her passion for research, writing, and history she agreed to take on the task of writing this book. Lois has a deep appreciation for war veterans, her father-in-law James E. Johnson Jr. was a World War II POW.

Lois and her husband, Jimmy moved to Weatherford in 1995 to be near their children James & Michelle Johnson, Sondra & Chad Floyd. Both children are employed in the Fort Worth area. Since then the family has grown to include four precious grandchildren, Chandler & Marley Floyd, Gracie & Gatlin Johnson.

Lois earned a Bachelor of Science and a Masters Degree in Counseling from Southeastern Oklahoma State University. She also received a Masters degree in Administration from Texas Women's University. She currently enjoys writing and teaching lessons for the ladies bible class at Lake Worth Baptist Church. Lois works for Millsap Independent School District, Millsap TX.

Mary Kemp

Thanks for the great cover design prepared by:
Photo Design Photography
Les and Cecilia Lopez
Coronado Ct. Weatherford, Texas
Long time friends who do great work.

Lieutenant William Lecel " Jerry" Lee

Dedication

This book is dedicated in Honor and Memory of Mary India Schoonover

Mary India and Leroy Schoonover

&

In honor of Leroy Schoonover and all Descendants of Lecel Lee 2009

Who Was Lecel Lee?

William Lecel Lee was born July 18, 1903, in Weatherford, Texas. He received a bachelor's degree from Texas A & M University in 1927, and was commissioned as a second lieutenant in the Army ROTC. After graduation, he went to work for the Purina Corporation but soon enlisted in the Army Air Corps. In 1935, after seven years in several Army Air Corps installations in Texas, Lee was ordered to the Philippines to assist with the establishment of a native air force. Shortly after his arrival in Manila, he became the Technical Advisor to the Philippine Air Force and was soon promoted to Chief of the Constabulary Air Corps. During the next three years he served as chief administrator of the Philippine military's air arm, which included overseeing a training program for Filipino pilots. It was during this time that Lee instructed Dwight D. Eisenhower to fly a Stearman Trainer.

On May 19, 1938, Lee's tour of duty in the Philippines ended. He returned to the United States and was once again stationed in San Antonio.

For most of the first year after his return, he traveled to aircraft factories throughout the United States to appraise planes for purchase by the Philippine Air Force.

In 1940, Lee was assigned to the Second, then the Twenty-Second, Bombardment Groups, Langley Field, Virginia. Lee spent the next year training pilots, performing various administrative duties, and conducting test flights of the new B-26 bomber.

In early 1942, Lee and his unit were transferred to Florence, then to Greenville, South Carolina. While there, he worked with both bomber and reconnaissance units. Later that year Lee was promoted to the rank of Colonel and was appointed training supervisor for the 21st, 320th, and 322nd Bombardment Groups at Mac Dill Field, Tampa, Florida.

In 1944 Lee was named commander of the 49th Heavy Bomber Wing and shipped to Bari, Italy, with the 49th and served as part of Nathan Twining's 15th Air Force. While in Italy, Lee participated in many bombing missions and was heavily decorated.

In the first half of 1945 Lee continued to take part in the Allied bombing campaign leading up to the defeat of Nazi Germany. Lee was promoted to Brigadier General and received the Distinguished Service Medal.

When the war ended, he was appointed deputy to Air Vice Marshall Brodie, chief of the Air Force sub-commission of Italy. He was assigned as commander of the Smokey Hill Air Force Base in Salina, Kansas, where he served from August 12, 1946 to September 15, 1947.

On September 15, 1947, Lee assumed command of Turner Field, Albany, Georgia and was there until August 1949. Perhaps his most notable experience while in Georgia was detail to the National War College from the fall of 1948 to the spring of 1949. During his tenure as commander of Turner Field he spent a large amount of time on tours around the continental United States during which he learned of new technologies relating to air power, such as atomic weaponry and jet aircraft.

After a short tour of duty at March Field, California, in the fall of 1949, Lee assumed command of the 3605th Navigation Training Wing, Ellington Air Force Base, Houston, Texas. Lee remained at Ellington until he assumed command of the Thirteenth Air Force, Manila, Philippines in September 1952.

General Lee returned to the United States as commander of an air force base near Amarillo, Texas.

He served as base commander until 1962 when he retired from the U.S. Air Force and joined the American Trust Life Insurance Company as senior vice president. Lecel Lee died February 26, 1976.

Compiled by: *Mary Kemp*
(From Eisenhower Library Material)

Lieutenant William L Lee reading to his son,
William L. Lee Jr. (Bill) while stationed
in the Philippines 1936

Forward
By
Jack Borden

The one thing that comes to mind when thinking about Lecel Lee is the compassion he had for people who could not defend themselves, or had someone imposing upon them. Lecel always volunteered, and tried to help. This is one incident that occurred when he was a senior at Weatherford High School.

There was a young lady who was also a senior and graduated at the age of fifteen. The teacher asked if there was a volunteer to put a geometry problem on the board. The young lady volunteered. The math teacher was a man. Well, the young lady put the problem on the board and the teacher looked at it and said, “that’s wrong, you didn’t work that problem correctly,” “Do I have another volunteer?” Lecel Lee volunteered. He went up and erased what was on the blackboard and redid the problem. When he was finished, the teacher said, “Now see that’s perfect- that’s the way I wanted it worked in the first place.” Mr. Lecel Lee looked at that man and told him, “You know as well as I do, that the way this young lady put the problem on the board was perfect. In fact, I copied what I put on the board from her paper. That’s the reason I know it’s correct. Now, you either give her an apology or there will be something else happen.” Well, the teacher, knowing how Lecel was accustom to taking over for the downtrodden, figured he would apologize.

That young lady, ended up to be my wife, Edith Jordan -Borden. We ended up living together as husband and wife for 66 years.

It was around the time of Lecel's death that she told me about how this had happened.

Another incident that I witnessed, where Lecel helped someone in need, was at a Mineral Wells and Weatherford High School Baseball game. During an at bat inning for Weatherford High School, about a half dozen Mineral Wells men jumped on Weatherford's star baseball player, and started beating him up. Lecel Lee was in attendance. He walked up to the bat rack - got himself a bat. When he got through, the five "fighters" were on the ground. The other one took off when he saw Lecel. This shows the type of man he was. He carried that compassion on through his life time.

In 1944, while Lee was in combat during World War II, his ability to show concern and compassion was revealed time and time again.

This is evident by the way in which his men talked about him. This statement is often seen regarding General Lecel Lee and his crew, "he is tough, but fair and compassionate." Major General Twining stated, " Jerry Lee is a fine combat officer, and if I had to go to war again, I'd like to have him with me." He also sent a report to the Pentagon regarding Lee's action in Italy during WWII.. Twining stated, "I consider Lee in the first eight of 116 general officers known to me."

General William L. Lee knew the importance of keeping the morale up while at war. I hear that he established "Lee's Poker Place" in the quarters for the 49th Bomb Wing. Each night the pilots gathered to play cards in an attempt to wear away the sadness and tears of war.

General Lecel Lee knew when to be serious about helping others. On several occurrences he would fly into combat with a group of new bomb crews who were inexperienced, in order to assist them while in flight.

Lecel also knew when to have fun and had a great sense of humor. He was notorious for pulling pranks on people for a laugh.

I recall Lecel's humor just after he got his flying wings. He flew to Weatherford from San Antonio. He talked his father into going up in the plane with him. Mr. Jeremiah Lee didn't want to go in that new fangled machine, but Lecel convinced him it would be an easy flight. " You'll just be up there floating in the air and nothing will happen." Well, he got the old man up there and he was enjoying it -suddenly Lecel rolled that plane on purpose. They were flying up side down. When they got to the ground, Mr. Lee told my dad, "Sam, that chew of tobacco came out of my mouth and hit me in the face three times before we got right side up!"

Jack Borden
Weatherford, Texas
December, 2008

Contents

Chapter 1

Early Childhood

Lecel Lee was born July 18, 1903 in Weatherford Texas to Jeremiah Linza "Jerry" and Rebecca Pool Lee. He was the first born, and only son. The family was later joined by the birth of Emily Ollie on September 1, 1904 and then Thelma Fay on October 21, 1907.

Jeremiah worked in the Weatherford Cotton Mill in Lecel's early years. Jeremiah wrote diaries from 1906 until 1913. Could this be because he was such a great friend of Bill Tanner who kept his diaries from 1902-1912?

Weatherford Cotton Mill November 14, 1903
Jeremiah is the tallest on the front row.

Lecel Lee
Lee Dairy Barn

Jeremiah Lee and Lecel Lee

Town Picnic of First Baptist Church Sunday School, July 4, 1914

1. Jeremiah Lee, 2. Josie Lee 3. Mrs. Will Tanner 4. Emily Lee 5. Thelma Lee 6. Lecel Lee

Thelma Lee- Hall

Lecel Lee

Emily Lee
–Tanner

May 29,1965

Tom Tanner
Husband of Emily Lee

Jeremiah Lee and
Thelma Lee

Josie Rebecca holding Bill (Lecel Lee Jr.)

Jeremiah Lee, Bill and Josie Rebecca Lee

Emily Lee Tanner

Jeremiah L. Lee

The Jeremiah Lee Family
Thelma, Jerry, Emily, Joel Rebecca and Lecel

Lee Family
Thelma and Bill Hall, Emily Tanner, Lecel and Bert Lee

Mr. **J.L Lee Died at Home East of City**

Born Dec. 29 1872
Died 1934

J.L. (Jerry) Lee, age 62 years, well-known dairyman of this city, died at his home one mile east of Weatherford at 7:30 o'clock Thursday morning, after an illness that had extended over a period of a year and a half or two years. During this time, his condition on several occasions became critical but for several months he had improved to such an extent that he was able to be up and about though still under the care of doctors. He was down town this week and made a trip to Mineral Wells Thursday, but was taken critically ill during the night and passed away at the hour stated above.

J. L. Lee was born at Carnesville, GA., Dec. 20,1871, and came to Texas with his parents in 1881. They settled near Hillsboro and Mr. Lee was there married to Miss Rebecca Pool in 1900. They moved to Weatherford in 1902 at which time Mr. Lee assumed the Manager-ship of the Weatherford Yarn Mill and he continued in that capacity as long as the mill was operated. Afterward he engaged in the dairy business and built up one of the finest Jersey herds in this part of the State. This business was continued up to the time of his death.

Mr. Lee is survived by his wife and three children, one son and two daughters as follows: Lieut. W. L. Lee, San Antonio; Mrs. B. H. Moorehead, Weatherford; and Mrs. Will Hall, Port Arthur, Texas.

The funeral services were held at the White Funeral Home Saturday afternoon at 5:00 o'clock, with Rev. E.S. Barlow, rector at All Saints Episcopal Church officiating. Burial took place in the city cemetery.

Mrs. J.L. Lee Passes Away Wednesday Afternoon

Joel Rebecca (Poole) Lee
Born 10-31-1862
Died 1937
Married: J.L. Lee in 1900

Mrs. J.L. Lee age 74 years, of this city died Wednesday afternoon at 6 O'clock after a critical illness that had lasted for three weeks. Previously she had been in failing health for some four years.

Mrs. Lee was born in Georgia and came to Hill County Texas with her family in 1900. In 1903 they moved to Parker County and to Weatherford, which had since been her home.

Three years ago. Mrs. Lee's husband the late J. L. Lee passed away and there are surviving two daughters and one son as follows: Mrs B. H. Moorehead and Mrs. M.G. Hall of Weatherford; Lieut. W.L. Lee of Manila P.I. She is also survived by one sister, Mrs. Mollie Black of Montrose, Ga.

The funeral services will be held on Friday Morning at 10 o'clock from the White Funeral Home, with Rev. Chas T. Whaley, Pastor of the First Baptist Church of Weatherford officiating. Interment will take place in Oakland cemetery.

The following will be pallbearers: C.A. Waller, W.D. Newberry, W.H. Wren, Rupert Kindel, Geo. McCall and Geo. Fant.

Chapter 2

Jerry Lee Family

Jerry Lee had one of the first and finest dairies in Parker County. After J.L. Lee's death in 1934, Lecel, along with other family members and employees worked to deliver milk.

Hubert Graft, Leon Tanner, Ruby Dunn and Emily Lee Tanner – Getting ready to deliver milk for the Lee Dairy

Old Bottle Cap

Leon Tanner in the pickup when he was a delivery boy. Britt Wren filling the tank of the Lee's Dairy

The Armstrong Family
Jerry Lee Dairy Employees around 1937

The Lee Story Goes On

By Leon Tanner

Jerry Lee, one of the first dairymen in Parker County, was very big and looked very tough – and he was tough. One day as he ate his lunch at a restaurant on North Main St., a local gentleman watched with interest as he ate the plate lunch. When he finished eating, he took his large pocketknife out and started picking his teeth. Jerry then walked up to the owner, and said. You know...the other day I ate a lunch just like this in Fort Worth, and the owner had the nerve to charge me 35 cents. 'How much is this one?' Now the regular charge at this café was also 35 cents. But, the owner quickly said, "just 15 cents, sir!"

When I was a youngster, I was very scared of Mr. Lee. We lived on what was then Old Dicey Road, now Jameson St., just behind the old Crockett School. I always walked up the road to our pasture and watered the cows. One day, Mr. Jerry Lee, came along in an old Dodge truck he used for hauling milk and offered me a ride. I said, "no" because I was scared of him.

He said "git in" and I walked around the truck and got in! When I told him to stop, he made me crawl over him and get out on the driver's side. I can imagine how much he enjoyed telling my dad, Bill Tanner, how he had scared me. I was so scared I could hardly water the cows. In 1935 when the Milk Strike was on, it has been told many times, that Lee Dairy of Weatherford never missed a delivery of milk to Fort Worth.

When one of Lee's trucks was stopped by the strikers who intended to pour the milk on the ground, the driver got out of the truck with his shotgun in his hand. He then told them, to go ahead and pour the milk out.

But, said they would then fall in it. They backed off and he took his milk to market.

About 5 or 6 months before Jerry Lee died, he and my father were drinking on our back porch in the afternoon. I was about 8 years old. I heard this story myself. They both made a vow that whichever one died first the other would ride their horse to the funeral. When Mr. Lee died everyone was upset because my dad, Bill Tanner, was not at the church service.

When we arrived at the cemetery on the north side of Weatherford, dad had tied his horse to a tree and was standing by the grave. The rest of the vow was that the surviving one would go to the cemetery the night after the burial and wet on the grave. My mother told many times that my dad got up in the middle of the night, saddled his horse and rode off. She knew he had completed the vow.

Leon Tanner

Jeremiah Lee

Bill Tanner

Milk Producers' Strike

The Daily Herald, Weatherford, TX, reported on March 26, 1935 that Parker County Milk producers joined with six other counties of the Dallas milk-shed area in a strike. Local dairies were admonished to withhold milk deliveries to distributors Monday morning. The dairymen were asking for an increase of price as well as an increase in quota on which the base price was paid.

Mr. A.A. Patrick stated in an unpublished article that this first strike of Parker county dairymen, did not involve probably more than a third of the milk production of the county. Several producers continued to sell their milk. According to personal reports the Lee Dairy was one that continued to sell milk and had been approached by strikers who intended to pour the milk on the ground. But, Jerimiah Lee did not allow them to intimidate him into surrendering the supply of milk.

Later an article in the local paper cited R.P Pittinger, President of the Parker County Dairy Association as having said that the reported rowdyism and destructive tactics in some counties of the organization had been exaggerated and did not reflect the "true situation".

An article from the Daily Herald found in the memoirs of the Lee Family which was dated 3-27-1935, indicated that no violence had been reported in Parker County, so far as was known and that not one gallon of milk had been dumped, but the strike continued.

The following night on March 30, 1935, the dairymen held a meeting and agreed on a declaration to end the strike.

According to reports the strike was not one hundred percent effective in any of the seven counties. A large loss was entailed both on the part of the producers and the distributors. The question was asked: Would the dairymen recover from the hard hit of the cost of feed and the loss of income from the strike? Only time would tell.

While the Lee Family was dealing with the issues of the dairy, General Lecel Lee was in the Philippines and had been charged with organizing and training the first Philippine Air Force. It was evident that both Lecel Lee, his father and family were brave and courageous people who were willing to take a stand even when the going got tough.

Ten years later, in 1945, the Parker County Dairy Association enrolled fifty new paid members at one of the largest gatherings of the year. Dr. Phil R. Simmons opened the meeting and H.O. Parnell read the minutes. He also declared that the organization was for the exchanging of ideas and improvement of the general production and marketing of dairy products.

C.O. Norton of Weatherford was elected President; I.C. Pittman, Weatherford Vice president; H.O. Parnell, secretary-treasurer; Roy Harberger, Aledo and U.G. Kelley, Weatherford as board of directors. The night was declared a success for Parker County Dairy owners.

My Memories of Uncle Lecel

By Ben Schoonover

It was a time before e-mail, the internet, cell phones, and the onslaught of mass communication. A Long distance phone call was a rarity and considered only when the message needed to be conveyed quickly or a glad moment in time when you hadn't heard the person's voice in a month or more. Back then, during the late sixties and early seventies, actual hand-written letters were the main means of keeping close contact with family and people you loved. I remember it being a great celebration when my grandmother, Emily Ollie Lee Tanner, better known to me as "Maw Maw", received a letter from her brother, William Lecel Lee.

These weren't just quick paragraphs that you ran a spell check on and hit the send button. But were well thought out words written carefully and precisely and carried much consideration which served a purpose. My small eyes compared them to a book and I remember Maw Maw reading them intently and by golly you didn't disturb her until she was finished reading.

We lived at my grandmother's house, which sits on the part of land still undeveloped where Uncle Lecel, Maw Maw, and Aunt Thelma, the youngest Lee sibling, grew up, so I had ample opportunities to view the remarkable things a hand written letter could do for a person.

My grandmother would scoff, say “um” and occasionally slap the many pages on her leg out of frustration and most of all, laugh, all in one letter!

They maintained close relationships through these letters, which held their brother/sister bond, a bond they cherished for their entire life.

I mentioned earlier that it was a celebration when a letter from Uncle Lecel arrived. It was a glorious day when he and Aunt Bert came by for a visit. Whenever their car pulled in the driveway, my two older sisters, Emilee and Becky, and I would race to see who would get to them first. Aunt Bert, who was always prim and proper, would daintily hug you and kiss you on the cheek, and was sure to leave the imprint of her always painted lips on your face. It was quite an amazing experience as a small child to be scooped up into the air and hugged and kissed by Uncle Lecel. It always seemed to be a mile in the air and to a small boy; he was close to a giant. Aunt Bert would always have a lovely smile on her face and Uncle Lecel would let out a boisterous laugh that would wash through your soul like a cleansing rain. When we were done with the hugging and greetings, Heidi, their faithful little dog that traveled with them, would amble out of the car with her tennis ball searching for someone to play fetch. She never had to search very long to find someone willing to throw it for her.

My memories of Uncle Lecel are few, but I do remember being on the listening end of many lectures. Sometimes they were things he thought I needed to know such as manners, etiquette, and proper behavior. I remember his booming voice that always had the sound of authority and confidence.

As he talked, each word was spoken with precision and elegance. My mom and dad, Mary India and Leroy, were always amazed that I wasn't afraid of him. He had a presence that demanded respect and an intimidating size that to some, could be overwhelming. I can tell you he had a ferocious thump which landed on my head more that a couple of times whenever my behavior didn't seem to be what he deemed appropriate.

I never really backed down but often found myself listening to the different methods I could have chosen rather than whatever it was I did to receive the thump in the first place. It always ended with a hug and then a kiss on the spot on my head that laid victim to the thump. I found that sitting in his lap was one of the safest places to be.

I never really knew the significance of his life when I was a small boy. Knowing what I know now, I would have been very inquisitive about his buddy, Dwight D. Eisenhower. I would have asked about his adventures in the Air Force and how he earned his wings. It never dawned on me that he had earned the privilege of being a Brigadier General. He could have shared his stories of crashing in the Philippines and explained to me how he became known as the, "toughest man in the Air Force." I never knew he had articles written about him in different publications. Even though he had many great accomplishments and was an outstanding leader of our great nation, to me he was and will always be, Uncle Lecel.

Ben Schoonover

The Schoonover Family

Parker County Courthouse
Weatherford, Texas

Chapter 3

Lecel's Education

William Lecel Lee received his formal education in the schools of Texas. A 1923 graduate of Weatherford High School, Lecel continued his higher education graduating from Texas A&M in 1927 with a bachelor of science degree, and was commissioned a second lieutenant in the Reserve on May 23, 1927.

Lee, W. L. *"Jerry"*
Weatherford, Texas
Age 24: 1st Lt. Artillery Staff

Trouble At Weatherford High School

School for Lecel Lee was not always smooth. During an interview with General Lee, he said, "I was a rough country guy, one time when I was sixteen years old, ex-President Taft came to town and there was a parade. I was a junior at the Weatherford High School, see, the juniors got into a fight with the seniors over a flag. It nearly broke up the parade. They suspended me. I took off out of town and chopped cotton in Arizona, went west to California, got a job milking twenty cows, picked apricots, and cultivated peanuts with two mules named Hank and Pete. Hank was a mean devil. Then time came that I wanted to join the Navy, but they wouldn't have me. They said, 'I was too young.' "

The question was asked, "Did you get kicked out of school more than once, General?" Lee's reply, "Well, yes, when my suspension was up the following September, I went back to Weatherford School. But then I got into a fight on campus and they suspended me again. They said they wouldn't suspend me, if I'd let the superintendent give me a licking, but I said, 'no'."

Once again young Lee took off to work, this time in Oklahoma. He worked on a bridge construction crew building a bridge across the South Canadian River. Finally a friend, John Mitchell, convinced Lecel to return to school when his suspension was up. He lasted out the final year and a half all right. Thanks to his friend John, he graduated in 1923.

THE MELON VINE 1923

Weatherford High School

Weatherford, Texas

Gladys Sears

" She counts her friends as riches, and is as rich as rich can be."

Ilene Bodiford

"My soul and body may be in Weatherford High, but my heart is not."

Lecel Lee

"Jerry"

" It's great to be on the inside-and it's safe."

Emily Lee

"My only regret is that I cannot give more years to basketball."

Sion Holley

" A man that could make Solomon take a rear seat."

WHS – Lecel Lee-1923

LECEL LEE

'19-'20-'21-'22.

As a tackle "Jerry" had no peer in Weatherford High. He was a hard man to meet on the field and he lived up to his motto of "Don't let them get up a-breathing." What will the team do without him next year. They will have to scrape to find another like him.

TEXAS A & M

Jerry chose Texas Agricultural and Mechanical College, the oldest of several state-supported colleges in the area.

When Jerry arrived he found it a beehive of 2,500 students. Lee stated, " The first year was rough. The upper class-men beat the hell out of me. I didn't like it, but I had to do something, so I stayed with it."

A reporter from the Saturday Evening Post asked Lee if he got into trouble in college. Lee replied, "I spent the last two years on probation, because I got liquored up at my cousin's wedding. She married Don Old (later Maj. Gen. William D. Old, of the U.S. Air Force). I was the only one that got caught by the night sergeant. They let me stay in school on condition I'd sign a letter of resignation they could keep handy. There was one other time, I thought President Bizzell was going to throw me out for milking a cow,"

Now the story goes like this, "A trainload of us had pulled up in a siding near Waco. A cow was tied up near the track, and I could see she needed milking. So I ran over and milked her. I didn't see anything wrong about that." Doctor Bizzell said, "It's the way you did it –out there before all those people." Jerry Lee continued at Texas A&M, until he finished in 1927. Lee graduated in the top 10 percent of his class of 322 students.

As far as we know Jerry Lee is the only person earning a bachelor of science, in animal husbandry to become a general officer in the United States Army Air Force.

THE LONGHORN

SADDLE AND SIRLOIN CLUB

LEE, W. L., *President*

RALPH, W. G. . . . . *Vice-President*
KINCAID, T. A. . . . . *Secretary*
MEY, J. S. . . . . *Treasurer*

CLASS 1927

AMEND, J. D.
FLOREY, A. J.
HOMANN, R. E.
MCCOY, V. O.
MILLER, J. H.
NEWSOM, F. N.
TODD, C. J.

APPLEMAN, J. C.
FLOYD, J. H.
LEE, W. L.
MCDONALD, C. W.
MILLER, C. A.
RALPH, W. G.

BLAIR, J. T.
HOBBS, L. E.
LUCKENBACH, F.
MATTHEWS, T. L.
NEWTON, R. J.
TATE, R. B.
WILLIAMS, H. L.

AMEND, APPLEMAN, BLAIR, FLOREY, FLOYD, HOBBS, HOMANN
LUCKENBACH, MCCOY, MCDONALD, NEWTON, NEWSOM, RALPH
TATE, TODD, AMEND, BIRD, BROWN, FARRIS, GAY

Page 353

1927

In 1927, W. L. Lee served as president of Texas A & M Saddle and Sirloin Club. The club is classified as an Academic-Agriculture Club which is recognized and attains national affiliation. The organization's purpose is to develop leaders in the Agricultural Industry and extend responsibility toward the community through special charities. The Texas A & M chapter was founded in 1920.

William Lecel Lee served as business manager for the Texas A&M Longhorn as well as president of the Saddle and Sirloin Club. The leadership of Lecel Lee abounds, despite the "rough first year." Throughout his lifetime Lecel Lee continued to "fight the odds." But,always seemed to bounce back and overcome obstacles.

After completing college, he signed up to go to Brooks Field, because he wanted, "to go down there and learn how to fly."

The problem was...he had to have 3 personal references. Now, that was somewhat a challenge to acquire. Despite the initial difficulty in obtaining the references, he was sworn into the United States Air Force in February 1928. Lee not only learned how to fly, but, taught many others to be pilots.

Honors and Awards: During Jerry Lee's School Years

Fish Football, 1923;

Distinguished Student, 1924;

Business Manager Rodeo, 1925;

Secretary-Treasurer Junior Class 1926;

Sergeant Major Artillery Battalion, 1926;

President Saddle-Sirloin Club, 1927;

Business Manager Longhorn 1927;

Vice-President Masonic Club, 1927.

Texas A & M, 1928.

Chapter 4

Marriage & Family

Lecel Lee married Bertha Ida Haag of San Antonio, Texas. They had one child – William Lecel (Bill) Jr. and three grandchildren: Karen, Karyl, and Jerry.

Brigadier General William Lecel Lee died on February 26, 1976 at the age of 72 years. He was buried in Fort Sam Houston National Cemetery, San Antonio, Texas on March 1,1976. He had been retired for fourteen (14) years and resided in Amarillo, Texas. His only child "Bill" was residing in England at the time of his father's death. The Diaries of Lecel Lee are archived in the Eisenhower Library in Abilene, Kansas.

Lecel, Bert and son Bill Lee

Bill Lee son of Lecel and Bert Lee

Bill Lee – " China Clipper" in the background.

Lecel, and Bert with son Bill

Lecel
Lee

Following in Father's footsteps.
Captain William Lecel Lee Jr. (Bill)
U.S. Air Force, Medical Corps

Family

Bert & Lecel Lee, photo taken on New Years morning. "We had to dress up to call on the president of the Commonwealth. Grandma Haag took the picture and tried to cut me out....I think" *Lecel Lee*

Christmas, 1965, San Antonio
Bert, Lecel and puppy

Lecel & Bert Lee in later years

Bert & Lecel with their Grandchildren

Back row: William Lecel Lee, Bertha Haag-Lee
Front row: Barbara Karen Lee, William L. "Jerry" Lee III and Brenda Karyl Lee. Photo taken at Edwards Air Force Base, California on September 6, 1964

Clark Field Entrance in the 1930's

Chapter 5

Military Service 1929-1938

William Lecel Lee was ordered to active duty at Brooks Field San Antonio, Texas, in March of 1928. He received his commission in the Regular Army after completing primary and basic flight training on June 1, 1929. This was the beginning of a career that took him across the world, and introduced him to some of the greatest leaders in military history. Lee was often asked how he, a rogue country boy who moved from rural Parker County Texas, became the "Founding Father" of the Philippine Air Force. Lee replied, "I was just a rough country guy, like my father. He was six feet two, weighed 230 lbs., and could pickup a 500 pound bale of cotton with two hooks and put it on his shoulder. I just came by it honestly."

The article in the Saturday Evening Post revealed many of Jerry Lee's rather rowdy experiences as a young man in high school and college. While attending school, he got involved in "little scrapes" and was "kicked out" several times.

Lee credits John Mitchell, a friend with which he worked while on a bridge building construction "gang" as being the one who talked him into returning to high school.

Lee said, "I never had any trouble with my studies, but I always got into fights. I was just a tough guy."

Jeremiah Lee, Lecel's father, finally put his foot down and presented an ultimatum to Jerry. He told his son, "you go to school now or not at all." Without hesitation, Jerry chose Texas A&M. This decision would set in motion experiences and opportunities of a lifetime.

Later, Jerry Lee received an appointment to the Naval Academy. But, chose to pass it up when he learned that he could not leave the Academy for the first two years.

He wound up working for Purina Mills, selling cow and chicken feed in various places in Texas. In a very short time he realized that he had no talent as a field salesman and returned home.

Jerry Lee found himself, "disinherited" after crashing his father's car. It was after the crash, while hanging out with a former Air Corps mechanic that Lee was encouraged to "just go down there at Brooks Field and learn how to fly."

William Lecel Lee left the life of milking cows, feeding chickens, chopping cotton, working as a feed salesman, and building bridges, to enter a new era and way of fighting and flying.

Lecel "Jerry" Lee was sworn into the United States Army Air Corps and began his primary training in 1929 at Brooks Field, Texas. He trained in PT-1 and PT-3 aircraft, then moved on to more advanced training in the De Havilland, (DH-4). Lee received flight instruction from two very prominent pilots Claire Chennault and Wendell Brookley.

In February of 1929, Lee was assigned to the 3rd Attack Group at Fort Crockett Galveston, Texas. It was here that he met and married Bertha, his wife.

The honeymoon was cut short because he soon found himself flying patrol in A-3 aircraft along the Mexican border out of Fort Huachuca, Arizona.

Additional instructor assignments took Lee to Brooks and Randolph Fields in Texas. Lee was one of the first officers stationed at the new training field at Randolph.

In 1935, Lt. William L. Lee received word that he was going to the Philippines. He looked over the information and decided he didn't want to go there! But soon learned that the choice was not that cut and dried. He encountered a superior who would "hold his feet to the fire" which meant he didn't have a choice. An irate lieutenant colonel called Lee's C.O. and said, "Lt. William L. Lee signed up for that job, and he is going to take it. Moreover, if he doesn't like the job he can resign in 'a couple of months'."

Lee arrived in the Philippines and found that the job was a real challenge. He decided to stick it out. This assignment led the way to Lee's many promotions, and gained him a place in history as, *"The Father of the Philippine Air Force."*

While in the Philippines, Lee was charged with organizing and training the first Philippine Air Force. His first duty was to draw up a list of regulations for the new Philippine Constabulary Air Corps. The airstrip had to be constructed and planes purchased for training.

Lieutenant William Lecel Lee met Ike Eisenhower on December 14, 1935. The purpose of this meeting was to discuss; the constabulary, building Camp Murphy, plans for the air unit, establishing physical standards for Filipino cadets, and the type of aircraft to purchase.

Major Eisenhower wanted information regarding Lee's interest in a squadron of airplanes for patrol of the coastline vs. Mac Arthur's plan for sea-patrol boats.

Primary Trainer (PT-1) used in the Philippines

Lee's diary entry stated, "When I arrived we had no men, no planes, no regulations, in fact we had nothing. It was two months before I received the first group of planes. What a beautiful sight, "three Stearman P-13's known to the Philippines as PT-1"

Lee brought the Philippine Air Force to life by carefully selecting students who became the first Air Corps' cadets. He spent hours training pilots, overseeing aircraft maintenance and building camaraderie among the people.

Lee developed a very good relationship with the Filipino natives, the officers and the training cadets. After his two year tour had ended, MacArthur wrote the war department requesting approval for a one-year extension of Lee's tour of duty. In his request, General Douglas MacArthur stated, "Lee has an outstanding fitness for the duties he is performing and his intimate experience with the Philippine army make him worthy of another year here."

The request was granted. Lee served in the Philippines from 1935 until 1938.

Lee was instrumental in having the new facility named Zablan Field after Major Porfirio Zablan. Major Zablan had been slated to become the Chief of the Philippine Constabulary Air Corps. Unfortunately, he was killed while undergoing training at Randolph Field. Lee felt it would be a fitting tribute to name the installation after him.

On May 2, 1936, the first aircraft flown over Zablan Field was a PT-3 piloted by none other than Lt. William Lecel Lee. This act ushered in the rebirth of military aviation in the country.

Zablan Field turned out to be the place where Lieutenant Lee and Lefty Parker experienced their most unique training opportunity. They soon became Lt. Col. Dwight D. Eisenhower's flight instructors.

Now everyone knew that MacArthur was not an air-minded man at that time and would have frowned on Lee and Ike's aeronautical training exercises. Actually neither Mamie nor General MacArthur knew that Ike was receiving flying lessons.

Colonel Ike just appeared at the airfield and stood around watching. Lefty and Jerry knew he had a keen interest in flying, so they initiated the conversation that soon became a common occurrence.

Lee's diary entry stated, "Ike never did ask us to teach him to fly. He would just show up at the airfield early every morning and we just took him in and started to teach him.

Of course, when he'd come out here and stand around, either Lefty or I gave him thirty or forty minutes time at the controls. Eventually, he got to where he could fly pretty well. We were riding up that morning and Lefty said, 'you know the old man could solo, but damned if I'm gonna let him do it.' Well, I said, why? Lefty said, 'well he might crack up.' I said, I'll ride with him and if I think he can solo, I'll let him do it. Because, I'm the boss out here."

"Ike took off and landed several times on the 'postage stamp' landing strip, and did quiet well." Jerry admitted, "if he had cracked up and got hurt or not hurt; I would have been in trouble. I'm sure Mamie would have been on my neck along with General MacArthur."

Ike flew the aircraft "beautifully" according to General Valdes, who was summoned by Lee to witness the event. Eisenhower encountered a problem with a sandbag ballast that was being used on the airplane.

General Valdes reported that the bag became lodged "between the rear cockpit stick and the rear seat, jamming the controls." But Ike managed to extricate the troublesome sandbag. "With quick thinking he maneuvered the plane in various positions, after which he landed and received a great deal of 'hand-shaking' and 'back-patting' for his professional handling of a difficult situation." This was indeed a happy day for the future President.

January 13,1937, "I rode with Lt. Col. Eisenhower for 50 minutes while he flew a PT-1. His flying was fairly good, but not as smooth as it could be."

PT-1 Stearman

January 14, 1937 "I had a new door put on the Stinson Reliant today, one that will come off by pulling a lever. This will allow the pilot to jump in case of an emergency."

Lee trained pilots and worked as a liaison between the Philippine Air Corps and the U.S. Army Air Corps. He flew a lot of trips back and forth for the officers of the PAAC and the USAAF.

"Today we left for Camp Murphy in the Stinson Reliant and landed at Baguio at 9:05. I was carrying two very prominent and important passengers, Col. Dwight D. Eisenhower and his son, John." Lee wrote about the incident in his diaries.

"Now this is where I liked to have cracked up, having Ike as a passenger in the Stinson Reliant. We left John in Baguio. He was returning to Brent from Christmas Holidays.

We stayed on the ground for about 50 minutes because the clouds were hanging around in the mountains. We finally left Baguio and landed at Camp Murphy at 11:25."

Lee stated, "I will never forget this. We landed all right when we got up there.

We landed uphill into the wind. See, the field was built on a slope. It's 5,000 feet or more above there, but we landed all right. The trouble came at take off. I decided we could take off downwind, down hill and go over a ravine; or we could take off up hill into the wind, and have to clear a big hump. We debated, we sat on the ground for almost an hour. I was just smoking my pipe and talking while we sat around the airplane.

Finally, we chose to take off up hill into the wind. We barely got over the ridge to the end of the field. I had that pipe in my mouth and Ike just sat with his arms across his chest. I said, Ike, those Bellancas, the Philippine Aero-Taxis, take off up this way with a load all the time. I think we can make it. Eisenhower replied, 'Well, you're the Doctor, and if you think we can, let's go.' I cranked up the plane and got around, squared away and took off. The progress was slow going up hill. I wasn't picking up much speed and definitely wasn't getting off the ground.

Like a darned fool and not thinking, I reached over and turned the flap lever, and of course it went all the way down. It was impossible to stop it half way. When the flaps went down, I kind of ballooned off the ground. But I did not pick up any forward speed. "Well, when I got up the hill, I just had nothing to do but try. When I got close to the edge of the hill, I pulled back on the wheel and we went over it.

On the other side of the hill was a 3,000 feet drop off. If we were not careful we could stall. When we went over, I just pushed that wheel forward as quick as I could.

We picked up enough speed and as soon as I could I let the flaps up, and we flew back."

Eisenhower stated in an article, "We didn't miss that hill but by a few inches." It was a sobering moment for Ike and Lee; one that both men remembered for some time.

On April 6, 1937, Lee and Eisenhower had another awful experience. He and Col. Eisenhower went to check on a cadet stationed at a camp in Tacloban. This time the weather was bad, but so was the bed!

Lee wrote, "Colonel Eisenhower, Cadet Zosa, Sgt. Antiago and I left Camp Murphy in the Stinson Reliant at 1:00 p.m. and landed at 4:15 p.m. While we were there inspecting the cadre, the weather turned bad. Once again we were grounded.

Eisenhower and I spent the night with Zosa's folks. Now there are good ways to sleep and difficult ways to sleep. That night we were introduced to the "Bejuca" bed.

Dr. Zosa had a nice house for the time. We had adjoining rooms, but the beds were just; well, nearly like sleeping on a board floor."

I said to Ike, I bet you're not going to sleep good on that bed, and he replied, 'Oh, hell, Bill, I'm pretty tough and I'm tired so I'll sleep.' Well now, Lee said, I didn't sleep very well myself, and I was reasonably certain that Ike didn't sleep either. When I got up the next morning, he was already up, running around bare-footed. I asked him how he slept. He replied, 'Bill, I'm not as tough as I thought I was,' Ike never said another word about sleeping on that Bejuca bed."

Lt. Lee continued, “That’s just the way he was, you know, he was just a whole soul type of fellow.”

Despite the near misses, Lee and Eisenhower continued to fly. Lee stated, “I needed the USAAF hours and Ike needed to inspect the training camps. So I flew him where he needed to go”.

In later years Lee reflected, “Sometimes it was really rough going down there. I’m telling you, when we were just topping the big trees and big, old high mountains around the islands. There were times that we didn’t know exactly where we were, and didn’t dare get on the instruments. We would turn around and go back or land somewhere and wait for better flying conditions.

We flew a lot of trips back and forth rather than driving. It was not safe to drive, especially at night. The streets were narrow and usually had local ox carts, as well as people on them. Given the risks involved, it was still faster and safer to fly.

We risked our lives for sure. That’s why I say, it’s amazing - the Good Lord was with all of us in our flying.

Like the times we were flying around in the unexplored area of Mindanao. That area was covered with trees 200-300 feet high. The weather conditions seldom allowed us to see more than 300-400 yards ahead of us. If we had crashed, we would have landed in those dense trees, never to be found. I never considered what we were doing as extraordinary at all.

It was just that the Good Lord was with us, no question about that. It was just the kind of things we did, just ‘normal business.’ We never thought about being afraid to fly.”

In the mean time, Lee and Parker continued Eisenhower's flying instructions. Lee's diary entry for May 19, 1937 stated, "I flew Colonel Eisenhower 20 minutes and then let him solo. This is the first solo flight he ever made and he was very happy."

Eisenhower's diary states, "It was on May 19,1937, that I soloed in a Stearman PT-1.

According to Lt. Lee, Eisenhower also learned to fly the Stinson Reliant, an O-19-C and a Stearman BT-1. Eisenhower was the first President licensed to pilot an airplane.

He was issued a private pilot's license. The license was granted through the Philippine Commonwealth. Listed on the application as instructors: Lt. William Lecel Lee, Capt. Mark Lewis, Lt. Hugh Parker, and Lt. Charles Anderson.

Eisenhower with the Philippine Air Corps Pilots USAF Photo
trained under Lt. William Lecel Lee.

Lecel and Crew Lost: Believed Dead

December started off with a whirlwind of trouble for Lee and his crew. The events during the week of December 4,1937, were printed in many newspaper articles across the nation and in the Philippines. Three planes, pilots and passengers are missing and believed to be dead.

Lee stated in his diary, "I took off from Camp Keithley with General Santos, Col. Segundo and Col. Stevens as passengers. We landed at Cagayan and took on some gas and distributed copilots and passengers among the three airplanes.

We all took off and headed for Cebu, but could not make it on account of the weather. We landed at Aureaguete and waited for the weather to clear. Bad weather kept hindering our take off. So we spent the night there.

The next day, the weather was pretty good flying weather in the Mindanaos, but really bad around Cebu. There was a report of a typhoon around Tayblos Island. We could not get to Manila, so we spent the night at Cebu.

Today the weather was fairly favorable, so we took off for Manila via Masbate. We had good flying weather the first hour and a half. It began to get worse and worse. It looked as if we might not make it. We were dodging thunderheads. When I thought it couldn't get worse, it did! I lost tabs on Santos and Franscisa.

I thought we could make it, until I had to start flying by instrument. We were lost at 10,000 feet altitude. We were afraid to come down, but we were running out of fuel."

Lee's diary entries for December 6-10 stated, "I told General Santos and Colonel Segundo to put on their life vests and they did. We finally sighted land and after maneuvering around some small islands, I picked out an agreeable looking beach.

We landed on the beach. The sand was soft and the plane nosed up, bending over one blade of the propeller. We landed unhurt and discovered there were two inhabitants on the island. They fixed us some rice, eggs ad fish. They allowed us to sleep in their nepa hut. What an experience, the worse I've ever had, but I guess it's all in a lifetime!"

The headlines were appearing everywhere:

Lieutenant W. Lecel Lee Reported Missing at Manila.

The Associated Press dispatch, received by The Herald reported that, "Lieut. W. Lecel Lee of Weatherford, Texas is missing along with four other army officers. The U. S. Army Air Corps and the Philippine Government have marshaled all available air, sea and land forces to search for the five missing men. Two planes with five men aboard had gone to the Island of Mindanao to investigate the Moro uprising in the Province of Lanao. Lee was piloting an army bomber and was accompanied by Major General Santos, Philippine Chief of Staff and Colonel Segundo, of the Philippine army intelligence corps.

1937 Philippine Centavo

CHINA
Taiwan
0 100 200 km
0 100 200 mi
Luzon Strait
Aparri
Luzon
South China Sea
Baguio
Philippine Sea
San Fernando
Quezon
MANILA
Batangas
Lee lost in this area
Legaspi
Mindoro
Samar
Masbate
Panay
Iloilo
Bacolod
Cebu
Leyte
Palawan
Guimaras Island
Puerto Princesa
Negros
Cagayan
Butuan
Sulu Sea
Iligan
Mindanao
Davao
Zamboanga
Jolo
MALAYSIA
Celebes Sea
INDONESIA
INDONESIA
Route due to typhoon
Regular route

Major Manuel Olympia and Lieut. Francisco are missing in another loaded army bomber.

The Philippine army ordered all available constabulary men to search the rugged mountain country. The United States Army prepared to send out a squadron of search planes at dawn. Government steamers in the general region were ordered to stand by to participate in the search if necessary.

"The hunt may be hampered by the typhoon, which turned from its northwest course and headed back toward the Pacific Ocean, leaving heavy crop and property damage and claiming several lives."

No Word from Lt. Lee; Weather Prevents Search

"A message today brings news that one of the lost planes made a safe landing at Infanta, Tayabas. The crew members reported that they had seen the other plane carrying the three officers. It left Cebu for Manila, but didn't land there. Around Manila there were reports saying that the plane had circled the storm and running out of fuel was forced to land."

Air, Sea, Land Forces To Hunt Missing Five

"Planes, ships and foot soldiers hunted today for General Santos, Chief of Staff on the Philippine army. He disappeared 48 hours ago in an army bomber with two other high ranking officers."

Fog Hampers Flier's Search

"The heavy fog thwarted the search for three of the highest officers in the Philippine Army. President Manuel Quezon, in a hospital recovering from an appendectomy, expressed apprehension over their fate.

Five planes returned from unsuccessful morning and afternoon flights through difficult conditions over mountainous Tayabas Province.

General Santos flew south to Lanao Province last week, when the Moro outlaws entrenched in cottas resisted efforts of insular troops to dislodge them.

The Moro chieftain reportedly swore a secret oath to take the life of the general and two other officers who attempted to persuade them to surrender their strongholds."

As the search continues, General Santos, Col. Segundo and Lt. Lee stayed around the nepa hut. Lee stated in his diary, "the weather is still rainy and there is a very strong wind blowing. Mr. and Mrs. Rafael Potestades have been very kind and accommodating.

The weather is clearing today. I took a bunch of pictures, both still and movie of the hut, Rafael and his wife as well as other interesting things this afternoon.

We heard an airplane about 5:00 p.m. and sighted it several miles to the southeast passing over another island of the Palillo group. It appeared to be a P-26 and was several thousand feet high headed toward Manila.

We drafted a message for Mr. Potestades to carry to Burdoes, then send it to Palillo for transmission to Manila. I think tomorrow will be good weather and we will be able to leave.

Colonel Segundo and I walked down to the plane, got our parachutes and some other things out of it, and locked the doors. We sighted an O-9 airplane several miles away. Infanta. I'm certainly glad to see you all are safe'."

It was about 25 minutes later when the O-9 # 7, flown by Lt. Fred Maxwell, sighted us and flew around to drop us the following message, 'Sales and Francisco landed safely near Infanta.

Lieut. Lecel Lee is Found; Safe on Little Island

"The first definite news of the safety of Lieutenant Lecel Lee, who has been missing since last Monday, was received in Weatherford, Texas on Friday."

An Associated Press dispatch dated Friday, stated that the missing men have been located on a tiny island where they landed when they encountered a typhoon. Relatives here were greatly relieved to receive the news.

Lee continued in his diary, "we were all glad to get home and see our families and friends. They were glad to see us back alive."

It didn't take the group long to recuperate and move on. Lee stated in his Dec. 11th writing, "I went to Nichols Field and talked to the ship captain, before he took off with Jew Lewis for Infanta in the amphibian, about the dismantling of the planes we left on Anirong Island and Infanta. They will be dismantled and shipped back to Manila."

Lecel Lee and Parker in the Philippines

Texan Rescued From Pacific Island

—Associated Press WIREPHOTO.

-Associated Press WIREPHOTO.

Manila, P.I., Dec. 16 - Lieutenant William Lee of Texas, Chief of the Philippine Army Air Corps, was rescued with two Philippine Officers from a tiny Pacific island after being stranded there four days when their plane was forced down by a typhoon. He is shown as the rescue ship reached Manila.

Gen. Santos, Col. Segundo and Lieut. Lee, Lost and Found

Details of the Airplane Mishap

By: Lt. W. L. Lee

Chief Philippine Army Air Corps.

This article was published in the KHAKI and RED Philippine Army Journal and General Magazine, December 1937.

The following radiogram was received at Cebu at 7:28 AM. "Number one typhoon signal hoisted three AM near Tablas Islands moving northwest or west by Weather Observatory, Manila."

The above telegram and weather report broadcast led me to believe that weather conditions were not favorable over the usual route flown from Cebu to Manila. However, it was believed that favorable flying conditions would be found over at the other route, which is from Cebu to Masbate and up to the Bondoc Peninsula. We had been delayed one day at Dunmaguete and one day at Cebu on account of weather. General Santos was suffering from an infection of the foot and all of us thought we should soon get to Manila. I was personally anxious to get back because I wanted to have an engine check made on our plane and return promptly to Cagayan to check the airplane which Lt. Andrews had damaged in landing.

In this connection it should be understood that, I, as the flight commander was solely responsible for determining when a flight should begin.

Neither General Santos nor any other high ranking Philippine Army officer urged me to start when I did. Believing that the weather was favorable for flying conditions, I made the decision to leave Cebu on the morning of December 6th. After flying the course for sometime the clouds became heavy and quite thick. The conditions to our west were much thicker and looked even more unfavorable.

So I turned into the fog to the east. After flying by instrument for quite sometime. I was not sure of our exact position. I thought we might be over mountains. I therefore decided, in order to play it safe, that I had better fly an easterly course for a few minutes.

I started losing altitude gradually in a big spiraling turn. After we were at about 6,000 feet, I advised General Santos and Col. Segundo to get the life vests from behind the back seat and blow them up by mouth and put them on. I asked him to blow one up for me. While he was doing that he saw a land formation to his left. I rapidly lost altitude down to approximately 2,000 feet so we could tell the heights of the hills, etc. After circling two islands we spotted a sandy beach on one and decided to land. General Santos wanted to know if we were going to jump. I said, “No” and that they should keep their safety belt on as tight as possible. I might nose the airplane up which could injure them. I made a wide circle over the water heading the plane into the wind, throttled down the engine and came in for a landing. The wind was very strong and the sand was very soft. When I landed the wheels sank into the sand some six to eight inches.

I did not roll more than 50-60 feet and just as the plane stopped rolling, it nosed up. One blade of the propeller was slightly bent and this was the only damage done to the plane.

I was up and down all that night worried about Lt. Sales and Lt. Francisco. As the leader of a flight I was responsible for the men that were flying the other planes. I felt that if the worst had happened I would be responsible for having led them to their death. I continued to worry. The next few days were very troublesome. Each day we looked at the weather conditions almost every hour on the hour.

On Friday morning we spotted an O-10 heading directly for the island. Shortly before reaching the island we heard the motor throttle down and the plane started on a long glide. We knew the pilot had spotted our plane. He flew around for several minutes and dropped a note which advised us that Lt. Sales and Lt. Francisco had landed safely at Infanta. Finally I could stop worrying.

Recently, I heard some discussion as to whether or not the Weather Bureau had rendered us efficient service. It did. In my years of flying in the Philippines, the Weather Bureau has been of the utmost help on all occasions, and I can say only that I hope it continues to render as good service in the future.

For Lt. Sales and Lt. Francisco, pilots of the other two planes this was an invaluable, never forgettable lesson. Certainly, it was for me. I must say, too, that it was the worst predicament that I have ever gotten into during my flying career. I shall not seek another similar one.

W.L. Lee,
1st. Lieut., A.C.

Jungle Strip Flier **By Art Brewster**

Chapter 6

Return to the U.S.
1939-1942

Lee's tour of duty was completed and he returned to the United States as operations officer at Kelly Field, Texas.

During an oral history interview, Lee recounted the event of his return to the U.S. on May 2, 1939. "We came home at the end of my tour of duty, which had been extended. Normally, tour of duty in the Philippines was two years at a time, but General Eisenhower, who was Colonel in 1938, and General MacArthur got me extended so I could continue my training of the Philippine Air Force." As the third year ended Ike talked to me about extending another year. But, the War Department in Washington did not approve another extension."

"So, I did not stay, I will never forget Ike, when he was telling me about this situation. He said, in a joking manner, "you're really an important man." He continued, "here's the former chief of staff of the Army wanting you to extend another year, and you are so important to the Air Corps that they won't let you."

At that time Bill was just a youngster, a little boy, and he was going to school. They finally agreed to extend me until the end of April, so he could finish school."

Upon his return to the United States, Lee was promoted to the rank of Captain.

Congress had passed a law that authorized the promotion of officers after ten years of regular duty.

Lee continued, “I did not know I was going to Kelly Field at that time. When I got there, they were not ready for me. They didn’t have anything to do that a corporal couldn’t have done. I wrote Ike about it, because when I got to Kelly Field, I was given the job of Field Maintenance Officer. It was an insignificant job, really!”

The problem it seemed was that the jobs assigned were not jobs that required a flying officer to handle. Again Lee replied, “It definitely did not require a flying officer. But their story to Ike and General Mac Arthur—as I remember, was that I was to head a special kind of a job to train a special type of student. Well, now, I wasn’t unhappy with the job I had to do”

Prior to this time, principal organizational change had taken place, the U.S. established the Air Crops as an equal combat arm, and made provisions for its enlargement and modernization.

Education and training began to receive greater attention as military officials reviewed the possibility of the impending great world conflict.

Army records show that the best training of larger units occurred in the Philippines, Panama, and Hawaii. The United States increased funding and reinstated airfields that had not been active since World War I. The need for more pilots, required more training programs in the United States.

Lee was assigned as base engineering officer for a while, then, in September 1939 he went to Maxwell Field, Alabama to attend the Air Force Tactical School.

After completion of the course Lee returned to Kelly Field and continued as base engineering officer.

Kelly Field, San Antonio, Texas was reinstated as an Advanced Flying School supporting the Army Air Force's effort during World War II. It was changed into a complex that modified and repaired aircraft, engines, and related equipment.

Lee wrote in his diary, "Well, when they reopened the field, they sent several officers to Kelly Field. Thomas Lynch, William H. Maverick, and I were sent there as part of the flight training program."

Captain Lee was soon moved to Brooke Field. Lee recounts a request that did not go through,

"A volunteer job came in, for a captain to go to the Philippines. Of course, I was one of the first to volunteer. A good friend, a classmate of mine from flying school, got the job. He had been back in the states from foreign duty longer than I had been back from the Philippines, so he got the job."

"Well, my friend, poor Bill, got to the Philippines in time for the Japs to hit the place. He was eventually murdered on a prison boat."

Lee stated, "Well, it could have happened to me. So I guess I was fortunate in that they didn't take me back to the Philippines at that time."

Lee did not stay at Brooks Field for a very long period of time, for in October of 1940, he was transferred to the Second Bombardment Group at Langley Field, Virginia. Lee served as squadron operations officer and was promoted to the rank of Major.

What did Major Lee do while at Langley? Lee said, "Well, I tell you what I did at Langley Field. That was when the 2nd Bomb Group had the B-18's. In those days we didn't have much of an Air Force, (Air Corps). We had the fighters, the bombers, and the attack."

B-18 Aircraft USAF Photo

The B-18 is practically the same as the old C-47. I was the squadron pilot in the 2nd Bomb Group. We had three squadrons in the group. I was the commander of the 96th Bomb Squadron, and Hal George was commander of the group.

"The B-18 is a twin-engine bomber. It had a shark's nose, it bulged out at the top and then dropped in the back" Lee said, "They had B-18's when I went there.

I didn't stay very long in the 2nd Bomb Group, things were, well, we'd already gotten into the War in England. We were anticipating it and we started expanding."

"The 22nd Bomb Group was organized and I was pulled out of the 2nd Bomb Group. Then a week later, I was assigned to the 38th Bomb Group, we ended up in Jackson, Mississippi.

During March of 1941, "I went to Wright Patterson Air Fore Base, Tampa, Florida, with eight other officers and three complete crews to run service tests on the Martin B-26. This was a newly developed type of aircraft."

"Lewis and I wound up as squadron commanders in the 38th Bomb Group. When we first got to Mac Dill Field, we didn't have any airplanes. Oh, we had two or three, old moth-eaten B-18's.

"It was like starting in the Philippine Islands." The group was there to run service tests on the new Martin B-26 aircraft and there weren't any planes to fly.

Martin B-26 aircraft *USAF Photo*

The Fort Worth Press article tells the story like this, "Tropical-hardened, rugged, square-jawed, Col. William L. Lee, commander of the new 21st medium bombardment group at Mac Dill Field is giving his fliers the same sure-fire offensive training he learned during his tour of duty in the Philippines under Gen. Douglas MacArthur."

In his diary entry in April of 1941, Lee stated, "At first we thought the B-26 was a pretty good airplane, but they were giving us some trouble. Matter of fact, they were popping rivets on the skin of the craft. Some had to be sent back to the Martin plant to have the rivets replaced. They were too modern for the landing fields of that time. The runways were too rough, the planes could not hold up."

Lee's observations and comments regarding the Martin-26 aircraft being tested proved to be very informative. He said, "I think the field technology of the retractable gear being used was partly responsible for failure of the landing gears.

One of the greatest areas of trouble came with inexperienced pilots or copilots pulling up, retracting the landing gear before the plane was airborne. Faulty cockpit procedure was the cause of several serious crack-ups.

The copilots would pull the gear up before the plane had the proper flying speed. The airplane must be running along at a high speed on the landing gear until speed allows you to pull the airplane up into the air.

If the landing gear is released before pulling up... well, one officer, P.D. Ent, died as a result of the landing gear."

"It was imperative to get the wheels up as soon as possible in those B-26's. If I had flying speed with that thing I wasn't afraid of it. The take off speed or lift off was usually around 100 mph. We landed it around 120 mph."

One newspaper article stated, "Lt. Col William L. Lee commander of the parent medium bomber group at Mac Dill, described it as a huge cigar with a couple of shingles stuck in the middle and rear." The most unique feature was the plastic nose which gave the bombardier unlimited vision, vital to accurate bombing.

Mac Dill field, largest bombardment air base in the southeast has been turned over exclusively for training of crews who will fly these great ships. The planes seen over Tampa today-may a week from now be bombing the Germans in Europe, or Africa, the Japs in Alaska or Australia."

Master Sgt. Raymond Harrell receives trophy from Co. W.L. Lee on behalf of the 398th Bomb Squadron for highest efficiency in the training program at MacDill field.

Lee described the B-26 as a "high shoulder wing, twin-engine, monoplane bomber very streamlined with a single vertical tail fin and slightly dihedral on the stabilizers on the wing. We used to call them, 'the flying prostitute' because there was no visible means of support."

He continued, "The plane had plenty of power. The problem was that they didn't have enough electric power to make the switch while landing, so they would over speed. They were powerful 2600 engines. That's when the term 'runaway prop' was coined. The tragedy is we probably lost an airplane at least once a week. When there was a crack up we lost everybody because they were too low to survive. There were no flight recorders or instruments that could help determine what happened."

Lee stated, "The most tragic B-26 crash was when Mark K. Lewis and Dave Laubach got killed at Briggs Field on the second day of the war, December 8,1941.

That crash killed seven of the crew of nine members. One officer and one enlisted man survived but were badly burned. Something had to be done."

Lee wrote, "It was about that time that I decided to write Ike and ask for help. I couldn't write General Arnold or anybody up there, 'cause I'd be shanghaied to the Aleutians' or some other place, I didn't want to go.

I knew Ike was in a better position to take it up with General Arnold, and some of the Washington people.

He did, the problem was looked into and a solution found. I don't think General Arnold ever knew where Ike got his information."

Once that electrical problem was solved, Lee stated, "I liked the plane all right, it made a good combat airplane. I have flown the B-26 many times, and despite the initial problems with the "runaway props", I can say it responds to the lightest touch. There is no plane better constructed, and she packs a tremendous bomb load for her size." The testing assignment on the B-26 aircraft was completed.

Lt. Col. Lee's next job was to train pilots in the B-25 aircraft. In early 1942, Lee arrived in Columbia, South Carolina where he was assigned to the 21st Bomb Group.

While in South Carolina, Lee's tough discipline once again created a "situation" with his superiors.

Quoting from his diary entry on May 1, 1942, "I wrote General Eisenhower a short letter today and enclosed a copy of the reprimand that I got from Major General Walter H. Frank yesterday, for making some of my soldiers carry folding cots as punishment for misconduct."

"Well" he said, "It happened like this." "We had these guys that would get drunk on payday and go *A.W.O.L.* for two or three days, then come back. I'd put them in *K.P,* but, they would break the dishes up.

So we made a "bullring" and they had to carry folding canvas cots for fifty minutes then rest ten minutes, as a consequence for breaking the dishes." Some one reported the action to the upper echelon. I received a reprimand for the action. My option was to accept the reprimand or accept a court-martial."

President Eisenhower rescinded the reprimand, and it was removed from his records, after the investigation was finished.

On November 9, 1942, it looks like U.S. Troops are about to take North Africa with Lieutenant General Ike Eisenhower as commander. Things were changing quickly. Colonel Lee was assigned to Meridian, then to Florence, South Carolina. His training skills were very important during the time of preparing for war. The United States was entering the Allied Forces of World War II. He was moved on to Greenville, South Carolina and assigned to the 49th Operational Training Wing.

Lee often seemed irritated by the fact that the upper command kept moving him around in order to train crews for combat. He wanted to be involved in combat.

Colonel Lecel Lee
1942

Lee wrote, "James E. Parker, who was commander of the 3rd Bomber Command called and they had received orders for the 49th Operational Training Wing to be converted to the 49th Bomb Wing. He wanted to know if I wanted to go overseas or stay. I told him, hell, yes, I've been wanting to go over since Pear Harbor." Colonel Lee's long awaited overseas assignment was coming to fruition.

In an article printed in the Weatherford paper in 1942, Lee stated, "I have a personal debt to pay the Japs for what they did to my American and Filipino buddies at Pearl Harbor."

Lee had his orders. In December of 1942, Colonel William Lecel Lee was chosen as one of the nine Air Force Officers to go to Italy under the auspices of the Fifteenth Air Force, commanded by Major General Nathan Twining.

During the period of 1938 to 1942, Lee was promoted from Lieutenant Lee in the Philippines, to Colonel Lee of the 49th Bomb Wing of the 15th Air Force.

Philippine Air Corps
Insigna

49th Bomb Wing
15th Air Force, Italy

The "China Clipper" on its first trip to Manila. It landed in Manila Bay, Philippines

Letter sent to Weatherford, Texas from Nichols Field Manila, P.I. Air Mail Via Clipper, Dec. 1935

Military Notes and Quotes

"Today, Col. Lecel Lee, wing commander of the 15th Air Force, stationed in Italy, was recently a guest of the Royal Household in Rumania, being entertained at lunch by the King and Dowager Queen of that liberated country. He could give no details of his mission."

Weatherford Democrat, 1942

"In 1952, Lee returned to the Philippines as Vice Commander of the 13th Air Force, The Filipinos were delighted to have their old friend back in their midst and Lee was once again swept up into an active social life as well as demanding military duties. When he left they were sad. It probably will be a long time before they again could enjoy a character as colorful and flamboyant

Saturday Evening Post

"Not only did General William Lecel "Jerry" Lee have the unique distinction of forming and training the first Philippine Constabulary Air Corps; he developed intimate and lasting friendships. This affinity for major events and people came naturally to the gregarious native Texas who, in addition to being highly dedicated, always performed his duties with character and humor."

Aerospace Vol. 20 #4

Chapter 7

World War II
1943-1945

While stationed at Greenville, South Carolina, the groups trained in the B-25 aircraft from April, 1941, to April, 1942, with anticipation of flying them in combat while serving in Europe.

Lee stated in his diary entry, March of 1943, "I didn't know until I got my orders for the 49th Bomb Wing, that we would be flying B-24 Liberators. That was disturbing since the pilots were well trained on B-25 aircraft.

I was sent to Italy as a wing commander, complete with staff officers, and everything. The biggest surprise came, when I got to Europe and didn't have airplanes or groups assigned to the Wing.

Now it's wartime, and we're just sitting there for two long months, waiting for the crews of the heavy bombardment squadrons."

Finally, the opportunity for which Lee had longed was taking place. In *April of 1943,* Colonel William L. Lee assumed full command of the 49th Bomb Wing in Italy. Lee received the staff and crews for the 461st Group, then the 451st Group, and finally the 484th Group.

The 49th Bombardment Wing still had decisions to make and work to do, when the groups arrived. Colonel William L. Lee and his staff members, Lt. Colonel Leroy L. Stefonowicz, A-3, and Major Arthur J. Lund, A-2, were busy visiting headquarters of the Fifteenth Air Force, and observing procedures in other Wings.

They had to select a building in which to house the 49th Bomb Wing Headquarters. The building selected had been used as a college of agriculture, and was located south of Foggia.

Colonel William L. Lee was now in Italy with his wing, as a Commanding officer capable of training pilots and leading groups into combat.

The 49th Bomb Wing was classified as heavy bombardment and flew B-24 Liberators. Colonel Lee wrote, "I didn't have but three groups, but I had one hundred eighty airplanes and about 900 people in the wing. You see, back then a wing commander was kind of a small air force guy."

"Being the junior wing commander in the 15th Air Force, well, the 49th would get the new groups. We would get the alternate groups that were always brand new crews with no combat experience,"

During a large majority of missions the wing had relatively inexperienced pilots and group commanders. This would test Colonel Lee's ability to train pilots in strategic attack, to apply theories and doctrines of flying planes in combat.

When the training program ended for the 461st Bombardment Group, Colonel Lee personally brought a field order to Colonel Glantzberg, CO of the 461st. T*he first mission was set for April 1, 1944.* "A feeling of expectancy filled the air and quipped the officers and men of the Group." It was time. It was time!

A report from the 461st gave details of the first mission. The target was the Bihac Marshalling Yard in Yugoslavia. "The mission was led by Col. Glantzberg, with 1st Lt. Joseph Donovan as copilot.

Capt. Marion Pruitt, Group Navigator was in the lead aircraft with 1st Lt. George V. Leffler, Group Bombardier. The Deputy Group Commander Lt. Col. Philip F. Hawes, Group Operations Officer Maj. William Burke and four Squadron Commanders; Maj. James E. Knapp, Maj. Robert E. Applegate, Capt. James C. Dooley and Capt. Edwin T. Grogan, flew the mission either leading a flight, or as Deputy leader in the No. 2 position in the "A" flight of each section. Capt. William Franklin and 1st Lt. William H. Tallant also flew on this mission."

The weather was clear and each crew flying on their first mission, watched the bombs hit the targets. The unthinkable happened as they watched the bombs drop, two aircraft collided over the target and were lost.

A USAF report stated, "the missions flown today encountered enemy fighters. Fierce air battles resulted in 19 bombers shot down and several missing."

Colonel William L. Lee now knew the feeling of loosing air ships and men during a combat mission. He related this in his 1944-1945 diary entries.

April 19, 1944, Lee wrote, "I drove down to the Stalian Hospital #61 at Foggia to see Col. Glantzberg the commanding officer of the 461st Group. He will probably be in the hospital for two more weeks. I drove to Costlluccio with Major White, to check out the problem they were having with the steel matting on the airstrip. It was cutting the aircraft tires as the planes took off and landed. We spoke with Colonel Eaton of the 451st Group and they are doing fine. Capt. John Mitchell, Pat Falley and I drove down to Torretta to check on the 484th Group.

April 20, 1944, Colonel Lee wrote, "I went on my first combat mission today. I rode as copilot with Lt. Warren. He flew in the lead position of "C" flight, in the second section of the 461st 36 Group formation. We were supposed to bomb the Marshalling Yards at Terarra, Italy. The target was overcast, so we bombed a railroad bridge and a road bridge at another place. We put in about 6 1/2 hours in the air. Saw some flak and a few enemy pursuit ships, *(aircraft)* but we didn't lose any ships."

A report from the USAF stated, "Strategic Operation, bombers again pound railway systems and road bridges in an effort to hinder the supply and withdrawal of enemy forces in Northern Italy. B-24s hit railroad bridges at several locations. Extensive damage was done to the rail system today."

April 22, 1944, "The weather cleared today and bombers resumed operations. The B-24's hit industrial areas at Schwechat and Bad Voslau as well as an air depot at Wiener-Newstadt. Enemy fighters attacked fiercely, downing 12 bombers,"

The 451st Bomb Group led for the first time by Captain Dooley, completely sprayed an enemy landing field with fragmentation bombs. The pattern was one of perfection which destroyed the targets." The group encountered a large number of enemy fighters. Fourteen aircraft were hit by flak, but there was no loss of life in the 49th Bomb Wing today."

The following communique was sent to the C.O.s of the 451st and 461st Bomb Groups. "The bombing of Bad Voslau, 23 April 1944, by the 49th Wing was excellent.Please pass to all participating my commendations for a job well done.

"The 484th Bombardment Group (Heavy) with B-24's became operational today. We now have 12 squadrons."

April 30, 1944, Today, the 49th Bomb Wing commanded by Col. Lecel Lee bombed bridges over Brenta River, hit supply dumps, highways, railway trains, and other targets.

May 5 & 6, 1944, The 15th Air Force's largest force to date, 640 bombers, attacked targets in Rumania and Yugoslavia. The B-24's hit the marshalling yards at Ploesti, troop concentrations at Podgoricu, Yugoslavia and an aircraft factory at Brasov. Major Knapp led the formation on the first mission for the 484th Group. One report stated, "Thirty enemy planes were seen, a few were encountered. Crew members were surprised at the amount of flak coming from guns placed in open fields outside the city limits. On the way back the formation passed over the defended Bor Mines area and was shot up badly by flak. As a result of the flak, the Group brought back its first man killed in action, 2nd Lt. Joseph F. Meyers, a bombardier."

May 17, 1944, Lee's diary entry for today stated, "I went on a mission with the 451st Group. I rode as copilot with the pilot of the lead ship, of the 3rd arc of the 2nd Attack Wave. Major Jaraby, Assistant Flight Surgeon went along in another ship. We bombed the Island of Elba and did a good job on the port installations and steel mill there. I flew for 5 hours and 30 minutes.

Today the attacks continued hitting the industrial area at Ploesti, Rumania, and targets in Yugoslavia. The weather was bad this week and several missions have been canceled"

May 18, 1944, Lee wrote, "I drove to Torretta Air Field to see how the boys at the 484th were doing. I also went to the 461st to see Col. Glantzberg for awhile. I presented Air Medals to Col Glantzberg, Lt. Col. Philip R. Hawes and several other boys today.

My groups went out today, but the weather caught them and they didn't bomb the primary target. They were able to hit the secondary site with success."

Colonel Lee pinning Air Medals on Col. Glantzberg, Lt. Col. Hawes, Col. Grogan, and Capt. Leffler 461st Bomb Group.

May 22, 1944, Lee's entry stated, " I rode with the 451st Group this morning on a combat mission in Northern Italy. I put in 6 hours and 10 minutes flying. The weather was not so bad today, but the primary target was partially clouded over. I don't know how the 451st Group did, as they were the only group that dropped on the primary target. The 461st and 484th dropped on secondary targets."

On the 24th of May, six heavy bombers were lost during enemy fighter opposition, while on a mission to bomb airfields at Bad Voslau. "Colonel Glantzberg was the wing leader on this mission. The cloud coverage was excessive. One crew experienced trouble when oil from a line on the nose turret guns froze, and obscured the vision of Captain Leffler, lead bombardier. Thirty enemy planes were encountered, twenty-three of our planes were hit by flak, and two were lost. One had an engine go out over the target, dropped out of the formation and was not seen again. This was a tough target."

Heavy bombardment continued through the month of May. The groups revisited tough targets like Ploesti, Bad Voslau, and Wiener-Neustadt several times until the target areas were rendered useless to the enemy.

The following memo was sent to the groups in the 49th Bomb Wing of the 15th Air Force, Italy

From: Lee, CO, 49th Bomb (H), APO52O.

To: Commanding Officer, 451st, 461st, 484th Bomb Groups

The following Teletype is quoted for your information,

"Cite FAF Baker 27, the terrific destruction inflicted on the enemy by units of the Fifteenth Air Force on the 29 & 30 May, is a record without precedence anywhere. Let us keep this high standard of attainment as our goal. The Hun can't take it!"

June continued to be an active month for the groups in the 49th Bomb Wing. Lee's diary entry for 5 June 1944, stated, "I flew on a combat mission over Northern Italy today with Capt. McClang as pilot in a B-24, of the 484th Group. We hit a railroad bridge. The photographs show that we did a good job."

Other reports stated, "Strategic operations hit marshalling yards and four railroad bridges today, carrying out a successful mission."

For the first time, the war in the air, being a new phenomenon was making a difference. The strategic operations of the 15th Air Force, were destroying airfields, bridges, railways and yards, oil refineries and other points that were vital to the enemy.

June 6, 1944, D-Day, Lee wrote, "The invasion of Europe by the Allied Armies of Europe started last night across the English Channel and from all news reports everything is going fine. My groups went after Ploesti again today. They had the place completely covered with a smoke screen and the boys had to bomb through the smoke. I don't know what results they had.

It was reported from the 484th Group that Operation FRANTIC continued today as B-24's hit Ploesti Oil refineries and a marshalling yard at Brosov. This is the day our ground forces invaded the coast of France from England. The B-24 bombers struck the German's source of oil at Ploesti with Capt. Edwin T. Grogan leading the mission. Grogan said, "It was 15 minutes of sheer terror."

The Ploesti target was not an easy area. It was an oil boom city in the plains below the Transylvanian Alps in the North. It is near the Romania capitol of Bucharest in the South.

USAF Photo 461st

B-24 Liberators of the 49th Bomb Wing, bombing Ploesti, Rumania The photo shows the screen of black smoke that billowed up from the smoke pots the Germans used in a effort to protect the oil refineries

Commercial refinement of oil began in Ploesti in 1851. Making it the first city in the world to tap into the riches of the earth. These resources became critical to the advancing technology of automobiles and airplanes.

By 1942, the refineries at Ploesti were producing nearly a million tons of oil a month, accounting for 40 percent of Rumania's total exports.

The Eighth Air Force's initial operation "Tidal Wave" was intended to take out the refineries at Ploesti. The 1943 mission was costly, 54 aircraft and 532 men were lost that day.

Later reports conclude that the United States Air Force did not have enough aircraft to do the job that would get the results, for which they had hoped. “Tidal Wave” inflicted damage, but the Germans quickly repaired the refineries and continued production.

Ploesti being a very large area required multiple missions to various parts of the compounds. In 1943, there weren’t enough aircraft to bombard the targets on a regular basis.

Most of the oil, as well as the highest-quality, ninety octane aviation fuel in Europe, went to the Axis war effort. Ploesti provided nearly a third of the petrol that fueled Hitler’s tanks, battleships, submarines and aircraft.

After “Tidal Wave”, Germany occupied Romania and protected her natural resources. More specifically, German gunners guarded the multiple refining and storage plants that ringed Ploesti. Antiaircraft guns, fighter squadrons as well as smoke screens were set up around the area to defend their valuable resource.

This protection set up by the Germans made it tough on the 15th Air Force. The 15th AF continued to bombard the area with heavy bombs, making at least twenty missions to different parts of Ploesti.

Colonel Lee’s diary entry for *June 7, 1944,* stated, “All three of my groups went out today, The 484th did a good job of bombing. I haven’t heard how the invasion of France is going other than it is progressing.

The 15th Air Force reached its planned operational strength of 21 heavy bomber groups, 7 fighter groups and 7 medium bomber groups.”

June 10, 1944, Diary entry, "I went on a mission with the 451st Group today and rode in the lead ship of the Second Attack Unit. We bombed oil storage installations on the East Coast of Northern Italy.

All three of my groups did a good job today. The 451st lost one ship that ditched off the East Coast of Italy and lost two of the crew members.

The other groups hit the Ploesti oil refinery again.

June 11, 1944, "All of my groups went on a mission today. I lost seven airplanes out of the Wing. Four out of the 484th groups, two out of the 461st group and one out of the 451st group. Four were lost to enemy fighters, two to flak, one crashed near home field and blew up. Another one was lost in landing, but none of that crew was hurt."

"Today B-24s attacked targets in Rumania and Yugoslavia, hitting oil installations. All groups had fighter escorts."

June 13,1944, Lee wrote, "Three groups went out on missions to the Munich area today. The 484th Groups lost six airplanes and the other two groups had no losses."

June 22, 1944, "I went on a combat mission with the 484th group today to Pola, Yugoslavia. I put in over 6 hours flying time as the Command Pilot. The weather was overcast at the primary targets today."

15th AAF reported, "Following five consecutive days of bad weather, B-24s bombed targets in Northern Italy, hitting six marshalling yards, two bridges, an automobile factory at Turin and an automobile depot at Chivasso. Fighters flew in support of the mission."

June 23, 1944, Lee wrote, "All of my groups went out on missions today and all did a good job." The B-24 bomber missions for today were oil targets in Rumania, oil refineries at Ploesti and oil storage at Guirgiu. We lost several planes and crews."

During the time between the end of June and mid July, the Strategic Operations of the 15th Air Force continued to gain propensity. The missions were becoming more intense and required more and more support by fighter squadrons.

July 4, 1944, Col. Lee's diary entry stated, "I flew down to Bari in the C-78 with Lt. Col. Lawhon as pilot and Lt. Col Nelson as a passenger this afternoon. We went there to attend a conference about a special mission coming up in a few days. Lt. Col. Stefonowicz and Lt. Col. Lund, others flew down in a B-25."

USAF Photo

General Nathan F. Twining 15th AAF Headquarters, Bari, Italy meeting with Wing Commanders: Lee, CO 49th, Acheson, CO 57th, Rush CO 47th, Upthegrove, CO 304th, Strother, CO 15th Fighter Control, and Lawernce.

July 5, 1944, "Today all three of my groups went on a mission to Southern France and all did a good job."

A USAF report stated, "Beziers Marshalling Yard, France proved to be a challenging mission for the Groups in the 49th Wing. The goal of hitting this target was to hamper the movement of German divisions from Southwest France to the front in Normandy. The photographs of this mission revealed that each group, 451st, 484th and 461st laid down almost identical bombing patterns. Later reports show that much damage had been done to the target."

"COMMENDATION"

From: Lee CO, 49th Bomb Wing (II) APO 520

To: Commanding Officer, 451st, 461st, 484th Bomb Group, APO 520 US Army

The following teletype is quoted for your information -FAF ASLE 72, Confidential:

"Your entire command is commended for the highly successful attack on German Troop trains in the Marshalling Yards at Beziers, this is one bunch of Huns that won't reach the beachhead."

On the lighter side, Lee wrote, "Lt. Col. Lund, Lt. Col. Stefonowicz and I went over to Foggia tonight to see Irving Berlin's "This Is The Army" as guest of Major C. O Holmgnew who commands the soldiers that played in it. I met Mr. Berlin and invited him out to have dinner with me Monday night. He accepted!"

July 6, 1944, Lee wrote, "I went on a mission today. The target was an oil storage plant in open country near an airdrome at the town of Aviano, Italy. It was a small target. Major Burke continues to be the Group leader.

I had Mr. Irvin Berlin out for dinner at the officers Mess tonight and showed him around my headquarters for a while, then drove him back to Foggia."

July 11,1944 diary entry, "Today, I got up at 5:10 a.m. and drove down to the Castellucio Airfield and went with the 451st Bomb Group on a mission to bomb the harbor at Toulon, France. I rode as Command Pilot with Lt. Tudor as pilot and Lt. Donahue as copilot. We flew for 7 hours and 20 minutes."

The USAF report stated, "Due to the weather, the only targets attacked today were at the harbor in Toulon, France. Eighty-seven B-24s hit jetties, oil storages, a nearby telegraph cable factory, barracks, repair shops, the adjoining marshalling yard and submarines in dry dock."

July 14, 1944, Lee stated, "All my groups went out today. The 461st and 484th did a great job, but the 451st had trouble. The mission to Korneuburg Oil Refinery, Austria lead by Lt. Colonel Hawes came through with a great percent of bombs dropped within range of the center of impact. There were encounters with enemy planes and lots of damage was done to our planes by flak."

It's time for the special mission that was planned several days ago. Tomorrow will be a test of the groups' abilities and determination to have success on a very important mission.

July 15, 1944, Lee wrote, "I got up at 5:00 a.m., and drove down to the Torretta, Airfield. I stopped by the 766th Squadron and had breakfast. Lt. Colonel Brooks Lawhon went with me. I went on a combat mission as Command pilot with Capt. Walter as pilot in the 461st Bomb Group. We went to bomb the Creditual Minier Refinery at Ploesti, Rumania. The mission took 8 hours of flying time. We got a good hit on the target. Lt. Col. Lawhon went with the 484th group and Lt. Col. Stefonowicz went with the 451st on the same missions. The 461st had one ship shot down over the target by flak."

The USAF report stated, "In Rumania, six hundred B-17s and B-24s bombed four oil refineries in the Ploesti area, and the Teleajenul pumping station. Records from the 461st and 484th reported, "the all out mission of the Fifteenth Air Force against the Ploesti Oil Refineries on the 15th of July had been designed as a final knock-out blow.

Earlier in July, the 15th Air Force held a conference with representatives of all the Groups to discuss this mission and to clarify pathfinder bombing. Specially selected crews received training in the pathfinder method and ran practice missions in other areas.

Colonel Glantzberg flew in the lead plane with him were the old reliable team of Captain Strong, Captain Leffler, Captain Pruitt, Lt. Sullivan and Lt. Gizelba. Colonel Lee flew as Wing Commander with Captain Waiters. The success of this mission earned for Colonel Lecel Lee and Lt. Gizelba the *Distinguished Flying Cross Medal.* Fourteen planes were damaged and one piloted by Lt. William L. Weems was shot down."

The missions continued, and according to reports Lt. Col. Hawes maintained his personally unpleasant tradition of being most unfortunate in drawing more that his share of rough targets. He often led formations through the toughest weather and was repeatedly attacked by fighters, and encountered targets obscured by clouds, smoke pots and haze. Despite all this, his determination trumped all odds. While en route to Manzell Dornierwerke Aircraft Components Factory in Germany, the group ran into weather that caused many of the pilots to return to their bases.

When the weather cleared, Lt. Col. Hawes rallied the wing formation for the bomb run. Despite heavy flak the mission was accomplished. The following communique was sent from Colonel William L. Lee.

From: Lee C. O. 49th Bomb Wing (H) APO 520

To: Commanding Officer, 451st, 461st, 484th

Bomb Group, APO 520, US Army

"Confidential, with reference to raid made 18, July 1944 and commendation received from Commanding General, 15th Air Force on same. I wish to add my hearty congratulations and deep appreciation for a job superiorly performed. This well performed mission indicates that it was expertly planned and excellent judgement on the part of the combat wing commander was used in its execution. It also indicated superior air discipline on the part of all groups and the personnel thereof. Please convey the above commendation to all personnel of your command and carry on smartly."

July 22, 1944, The 461st hit Ploesti again today. The groups were hit by heavy antiaircraft guns upon reaching the target. All the planes dropped their bombs as briefed. They fell short of the intended target. Of the twenty two planes on the bomb run, seventeen were hard hit by flak and four others were lost. Two planes piloted by men who were on their second mission left the formation and disappeared. Two crews parachuted out near base due to running out of fuel and loosing an engine.

By the end of July, most of the original combat crew members had completed their missions. Replacements were moving in, leaving the groups with inexperienced crews.

The new crews were often assigned "freshman missions" allowing them some practice before hitting targets that were heavily defended.

During the remainder of *July 1944*, Lee wrote, "I drove to Torretta Airfield this morning and went on a mission with the 461st Group. Each Squadron had individual targets. I went in the lead ship of the 764th Squadron. We bombed the town of Sjenica, Yugoslavia with 1000 lb. bombs. I think we did a good job. All flights hit their targets successfully. The 766th Squadron, had the largest target. They dropped 88 percent of the bombs on the aiming point. The mission on the whole was a good one."

"Today 25 July, all three of my groups went on a tough mission in Austria. The 461st group lost 12 ships to enemy fighters. A number of parachutes were seen to open and according to eye witnesses, the dammed German Fighters shot up the boys and the parachutes as they were coming down.. Major Burke, C.O of the 766th Squadron was lost. The 451st and 461st didn't lose any ships."

Colonel Lee, Wing Commander, experienced much grief regarding the mission to Herman Goering Tank Works, Austria. A report from the 461st history stated, "The target was heavily defended and the German fighters took advantage of the inexperience of the new crews. Most of the planes flown did not have their ball turret down on the bomb run, allowing twin engine planes to come up under the lead flight of the formation and throw rockets through the bomb bay doors. Major Burke's lead plane was the first to go down. For the first time in its history, enemy fighters successfully turned back the 461st Group short of its target."

Two days later Colonel Lee wrote, "I went on a combat mission, my 13th, with the 461st Group this morning to bomb the Marshalling Yards at Phlorina, Greece. Major Joe Juna went along too. The 461st did a good job today. The other two groups didn't have as much success at Ploesti."

August 1, 1944, " Today, Capt. Bill Booth was promoted to Major, Lt. Thackston and Lt. Wallace were promoted to Captain and 2nd Lt. Sees was promoted to 1st Lt.

The next day Lee wrote, "I went on a combat mission with the 451st Group today. We bombed the Le Pontet Oil Storage in Southern France. The ship I was in got about 10 holes shot in it at various places. This was the toughest mission I've been on."

Colonel Lee reported on the 7th, "Colonel Taylor flew up to my headquarters in General Twining's C-47 and picked me up. We flew to Rome to the U.S. Army Air Forces operated, Majestic Hotel. The rest and relaxation was needed.

"During the afternoon the "hot information" on the big push came down through channels. "H" hour of "D" day was to be the break of dawn on the morning of August 15th. The 15th Air Force had 4 objectives for the day: 1) To destroy enemy coast and beach defenses within the battle area; 2) To destroy the remaining rail and road bridges across the Rone River; 3) To block defiles and rail lines from running through the Alps southward; 4) To drop propaganda in the beach areas."

August 15, 1944, "The 461st Group participated in the second "D" Day operation with the bombing mission to the landing beaches in Southern France."

August 17 1944, Lee wrote, "All three of my groups went up to bomb the Refineries at Ploesti Rumania today and did a good job."

Commendation

From: Headquarters Fifteenth Air Force,

Office of the Commanding General, APO 520

To: All Units, Fifteenth Air Force, APO 520.

"1. The commanding General, United States Strategic Air Forces in Europe, has requested that all personnel of the Fifteenth Air Force be commended for their part in the operations of this Command with similar operations by the Eighth Air Force, have been most effective blow so far struck against German industry."

"2. It gives me much pleasure to pass on this tribute. I realize that our successes at Ploesti would not have been possible except for the skill and industry with which all of you, in both air and ground echelons, performed your assigned tasks. By our work you have made major contributions to an early defeat of the German war machine."

Signed N.F. Twining
Major General, USA

The groups in the 49th Bomb Wing continued flying missions almost daily. Lee's diary entries stated:

"All three of my groups went up and dropped frag bombs in Northern Yugoslavia and did a good job.

"Today, my three groups made and attack on the airdrome at Mardersdorf. The 451st lost nine ships to enemy fighters. This is the second day they have suffered heavy losses."

On the lighter side, "the heavy weight Prize Fighter, Sgt. Joe Louis, addressed the officers and men who had assembled in the Group Theatre to greet him. All had a good time."

"I made my 16th combat mission today with the 484th Group. Lt. Stone flew as pilot and I as command pilot. We attacked the railroad bridge across the Po River at Femara, Italy. My wing lost several ships, but all crews got out in parachutes. After returning from the mission, I visited with Col. Glantzberg.

Records show that the experience level of the combat crews during the month of August was the lowest in the history of the wing. Seven of the eight crews lost in the month of August had only flown a few missions.

Forty seven new complete crews were assigned to the 461st during the month. The commanders developed definite policy that was rigidly adhered too. It was intended to prepared the new crews as much as possible."

461st Base Camp, 1944

461st Headquarters

September brought about major changes in command for the Groups of the 49th. According to a narrative history report, Colonel Glantzberg flew his fiftieth sortie and in so doing, ended his tour of duty. Lieutenant Colonel Philip Hawes became the Group Commander.

The objective for the 461st Group changed from combat to that of supply transport. This unique opportunity allowed many pilots to fly at lower levels over previous targets. They got actual visuals of some of the damage afflicted during their most successful combat missions.

The other groups continued their bombardment activities disrupting railroad communications in Northern Italy.

Some of the targets hit by the 49th Bomb Wing during September included: Tank concentrations in S. Yugoslavia, Railroad bridges at Brod and Belgrade and marshalling yards at Sarajevo and Nich. They hit submarine bases in Greece, airfields in Tatoi, Eleusis and Kalamaki.

During the third week of September the transport missions were complete and the group converted back to a bomb group. They continued to participate in the attack on oil refineries, marshalling yards and bridges.

September 24, 1944, Lee's diary entry stated, "I went on my 23rd mission with the 451st Group today. We bombed the Athens/Eleusis Airdrome near Athens Greece with 1000 pound demolition bombs. The mission was accomplished. The bomb pattern started on the west edge of the dispersal area and continued south along the east side of the airdrome. Six enemy planes were destroyed." For the next few days the major targets were harbors in Greece.

Commendation

From; Headquarters Fifteenth Air Force, APO 520 US Army
To: All units in this Command

"The air support evidence in the invasion of Southern France is an unsurpassed example of the role to be played by the Air Forces in an operation. The fighting spirit and determined will to crush the enemy demonstrated by every individual regardless of rank or position, coupled with the brilliant leadership displayed, have won the admiration and gratitude of the entire nation. When "The Yanks Are Coming" the issue is not in doubt. Please convey this message to all participating units."

By Command of Major General Twining

Forty-ninth Wing history stated, "the last mission in September, by the 49th Bomb Wing was to attack submarines in the Athens, Greece area. The only crew lost to combat during the month of September was lost on this mission. Second Lieutenant Ralph E. Newton, left the formation before reaching the target but failed to return to base. His crew was reported as Missing In Action."

October brought on colder weather and an order from Colonel Lee. "Woolens have replaced khaki as the proper uniform. In anticipation of winter, the tent floors had to be covered with brick, tile, lumber or gravel and the tent walls needed to be wood or tofu of at least 1 1/2 feet high. The crews started using their ingenuity to prepare for the hard winter. Frag bomb boxes began to appear as floors and walls in tents. The men had resigned to the fact that they would indeed spend the hard winter in Italy. They began to take steps to make themselves as comfortable as possible.

Winter camp for one of the 49th Bomb Wing Groups, Italy, 1944

The average temperature in the aircraft during this time was as much as fifty-four degrees below zero. The men had to be prepared for the weather and altitude over the Alps.

Lee's diary entry *October 6, 1944*, "We took off in a B-17 with Col. Barrow as pilot, Bob Acheson and I and several other passengers. Two, B-17 with General Chancey and Colonel Batjez as passengers made up a flight of three. Eight P-38s joined us for fighter escort. On our way to Bucharest. We were shot at by German flak unexpectedly while flying at 14,000 feet over Yugoslavia. We were lucky that none of us were shot down because the flak was intense accurate and heavy. While in Bucharest we enjoyed a good dinner of soup, fish, steak and ice cream.

The next day "Lt. Col. Eaker, Bob Acheson, Mr. Vasilescu and I drove out to Ploesti about 40 miles away from Bucharest and looked at the damage to the Americano Romania Refinery we hit when we bombed the place several times. It was destroyed."

On the 8th, Lee wrote, “Several of us drove up to the Kings Summer Palace at Sinaia to meet and call on the King, who is a boy of 23 years old and the Queen, his mother. They were both very nice and talked excellent English. We were shown through the palace by an attendant. It is the most elegantly furnished place I have ever seen! There is no question but what the furnishings cost several millions of dollars.

Coming back from Bucharest we ran into some artificial smoke at Ploesti and it was just like heavy fog. It was very effective in creating a black haze. It is the kind they used to put up smoke screens when we bombed Ploesti.”

“I am impressed with Romania and it’s people. It is very pretty farming country and all the people are industrious. They are busy repairing the bomb damage done to their towns and plan to build it back as soon as possible and do a better job.”

“We planned to take off for Bari, Italy, today. When our P-38 escort arrived they told us that the weather wasn’t very good. We decided we would not try to make it. We finally got to take off the next day. We got things together and loaded an injured enlisted man aboard General Chancey’s ship. The man was shot down over Ploesti on August 10th and has been in the hospital ever since, and will be for some time to come. He is the only survivor from his crew.

November was even more difficult, once again the weather became a huge factor in the number of missions flown or completed.

A narrative report from the November history of the 461st stated, “The outstanding event of the month of November for all members of the Group was the awarding of a Department Unit Citation to the Group by Major General Nathan F. Twining.

Major General Twining's presentation to the 461st and 484th Bomb Groups, Bari, Italy

Commanding General of the Fifteenth Air Force
This message came to the headquarters on:

November 6, 1944

Commendation Unit Citation

The citation was received by the unit "For outstanding performance of duty in armed conflict with the enemy." The citation specifically mentioned the successful destruction of the Duna Repulogepgyar Aircraft Components Plant at Budapest, Hungry. The report praised the ground crews for their diligence and arduous work to have their aircraft in the best mechanical condition to insure the success of this attack. Thirty eight B-24 bombers set out for the target and despite severe and adverse weather, maintained their formation. The Group which had no fighter escort, encountered enemy fighters firing rockets, cannon, and machine guns in an effort to bust up the formation. The crews continued with courage and determination to hold the formation tight and battled their way through the enemy defences and accomplished a highly successful bomb run. The group destroyed 3 enemy fighters and held their losses to 3 bombers.

Major General Twining wrote, "by the conspicuous gallantry, determination and professional skill of the combat crews, together with the superior technical skill and dèvotion to duty of the ground personnel, the 461st Group has reflected great credit upon itself and the Armed Forces of the United States of America."

The Unit Citation award was presented by Maj. Gen. Twining, Colonel William L. Lee, C.O. 15th Air Force, 49th Bomb Wing Commander, Lieutenant Col. Hawes, Commander of the 461st Group and Lieutenant Col. Bush, Commander of the 484th Group.

The groups in the 49th continued to beat all odds, against the weather and enemy defences into the month of November.

Commendation

November 19, 1944

From: Lee, CO 49th Bomb Wing
To: Commanding Officer 461st Bomb Group

"The following teletype is quoted for your information. With pride I pass to you and your officers and men the following cable from General Arnold."

"Fifteenth Air Force, operations from 16-19 November, 1944 have been noted with appreciative interest. Worthy of commendation is the sustained effort on successive days. Convey to all members of your command participating, my heartiest appreciation for a job well done.

The ground maintenance crews in particular should be most emphatically informed that their untiring efforts are most deeply appreciated not only by air crews, but by all member of higher echelons, especially myself. The AAF is indeed proud of the men and her ground crews who make possible each new air success."

General Arnold

November 22, 1944, Lee wrote, "I worked around my office today. I got a message from General twining tonight telling me that I had been nominated for the grade of Brigadier General by the President of the United States. If the Lavate approves it, I should get my promotion in a few days."

On the lighter side, " Mr. LaMarc the Italian who managed the place where my headquarters are located, gave me a live turkey for Thanksgiving today. I put him in a pen to feed for several days before we kill him to eat."

December ushered in a different type of problem. The weather was extremely difficult and often set the pace for missions. There was snow in Italy and extreme weather over the target areas. According to a report from the 451st Group St. Sgt. William Smith, lower ball turrent gunner in a B-24 bomber, kept passing out from the -54 degrees temperature during a mission.

December 7, 1944, "Only pathfinder planes went out today. Local weather was partly cloudy and cold all day. I had a message from 15th Air Force to be at the headquarters tomorrow for a Wing Commanders's meeting at 10:30 a.m.

General Twining, 15th AAF Headquarters, Bari, Italy meeting with Wing Commanders: "Jerry". Lee, 49th, Acheson 57th, Rush 47th, Upthegrove 304th, Strother 15th Fighter Control and Lawrence 5th Wing.

"I had a call from Major General Twining about 4:30 today, telling me that I was promoted to the grade of Brigadier General by paragraph #1 war department Special Orders No 293 dated 7 December, 1944."

The Fort Worth Press

"Weatherford Flier Goes Up to Brigadier General"

Word that Brig. Gen. William L. Lee, native of Weatherford and graduate of Texas A&M College, has been elevated to the rank from colonel, reached him in Italy recently while his B-24 liberators were approaching their 150th mission against enemy targets in Europe.

The flying general, heads the youngest wing in the 15th AAF but one that has pounded all of the major objectives in the German network of factories, oil fields, and communications systems in its eight months of operation.

Under his leadership the groups in his wing, have taken part in raids, on Ploesti oil fields in Romania, the aircraft industry at Winer -Neustadt, the synthetic petroleum plants at Blechhammer and Vienna, airfields in Austria and Balkans, and have been cited six times by the President of the United States and the War Department.

General Lee received the Distinguished Flying Cross for superior leadership on a concentrated bombing mission against a strongly defended target in Romania on July 15, 1944. He organized his present command and took it overseas last January.

Weatherford Democrat

"Parker County is Boasting of Four Generals Now in Active Service"

Parker County can now boast of four generals that are serving in World War No. 2, the latest being added to the list, when Col. John T. Walker of Azle, was elevated to the rank of Brigadier General a few days ago. The four are:

Lt. General Hood Simpson

Brigadier General Lecel Lee

Brigadier General Homer Sanders

Brigadier General John T. Walker

General Simpson is in command of the 9th Army now advancing into Germany.

General Lecel Lee is wing Commander of the Air Force that is now stationed in Italy. He was Captain when the war started and his promotions have been made regularly and consistently.

General Homer Sanders is command of a wing in the Air Force that is serving the India-Burma theatre and has been there for some time. He was reared and educated in the Pythian Home of Weatherford.

General John T. Walker is in the Marines and as Colonel of his Regiment, has made history in the invasion of Eniwetol Atol and at Guam. He was reared at Azle, where his father J. R. Walker was druggist for 43 years."

Weatherford Paper, 1945

"Hello, General" Over Phone Tells Colonel of Promotion

The greeting, "hello, general" on the telephone was the first time Col. William L. Lee of Weatherford heard that he had been promoted to star rank in the Army Air Forces, according to a communication from the 15th Air Force in Italy...."

December 17, 1944, Lee wrote, "All three of my groups got jumped by German fighters today. The 461st lost ten ships, the 451st lost four and the 484th lost two."

A report from the 461st stated, "Captain Mixon took off with thirty one airplanes to attack the synthetic oil refinery at Odertal, Germany. As they neared Muglitz, just south of the initial point, the group was attacked by as many as fifty Me-109s and FW-190s.

The attack lasted about fifteen minutes during which the enemy fired rockets and 20mm cannons against the group. As a result of the encounter with the enemy fighters, sixteen B-24s were lost, 3 men killed and 93 men missing in action.

To make matters worse, a German voice, using our radio frequency, broke in and asked, 'Where is the rest of your formation?' laughed and signed off. The loss was great and the insult was one that would never be forgotten."

The remainder of December was difficult. Most of the bombing was done by the pathfinder method because of thick cloud coverage.

According to group reports there were several breaks in the coverage and photos revealed exceptionally well hit targets.

Commendation

From: 49th Bomb Wing, General W.L. Lee
To: 451st, 461st & 484th Bomb Groups
Attention: S-3 Officers

Recent photographs of coverage has proven that our relentless attacks have dealt crippling blows to the Hun and his most vulnerable spot-his sources of oil. Definite information verifying the effectiveness of the tremendous effort the combat and service units of this Air Force have been called upon the exert compensated in measure for their sacrifices entailed. I realize that in ordering deep perpetrations into enemy territory with unfavorable weather adding to the hazards of combat, I have required the personnel of this command to exert a supreme effort. It is heartening to know that such an effort has been crowned with success.

The devastated refinery of Blechhammer north bears witness to the valor, the outstanding flying proficiency of professional skill of our combat crews and to the determination of their combat leaders. I commend the fighter pilots and their leaders on the brilliant performance of their escort assignments.

I congratulate the service units and the maintenance personnel, the tactical units on the excellent response and arduous demands made upon them. We know that our campaign is progressing favorably and that our successful attacks on the enemy in the recent months will hasten his complete and utter defeat."

Signed: Major General Nathan F. Twining

During the days between Christmas and New Years, B-24 Liberators from the 49th Wing flew missions to three distinct targets in Italy and Poland. The group performed with such accuracy that the wing earned three commendations from General Twining.

Two missions were flown against the Venzone Viaduct in Italy. Photo reconnaissance revealed the wing had made successful runs and had indeed cut the target in half.

Commendation

From; C. O. 49th Bomb Wing Command Section
To: C. O. 461st Bomb Group

"The following message from General Twining is quoted for your information, 'My congratulations to the 49th Wing for superior bridge busting job at Venzone on the 27th and 28th of December. Please pass to all Groups participating Well Done'."

General William L. Lee

For the remainder of December, heavy bombers from the 49th Wing flew strategic missions in the Vienna area and rail points carrying supplies. Escorted by Lightening and Mustang fighters, Liberators bombed three targets on Brenner route from Munich into north Italy and three rail yards in Vienna.

These targets were main lateral lines of supplies for the German's SW Front. The target objective was to destroy enemy military supplies and create blocks on both sides of the trapped Germans. Mission accomplished was the message of the day!

Commendation

Subject: TWX 00504
To: All Concerned

"The following message received from General Twining is quoted for your information. The results of the bombing by the 461st group against the alternate targets, the locomotive depot at Castelfranco, were most excellent. My highest commendations to the crews who turned in this splendid performance."

Signed: General Lee

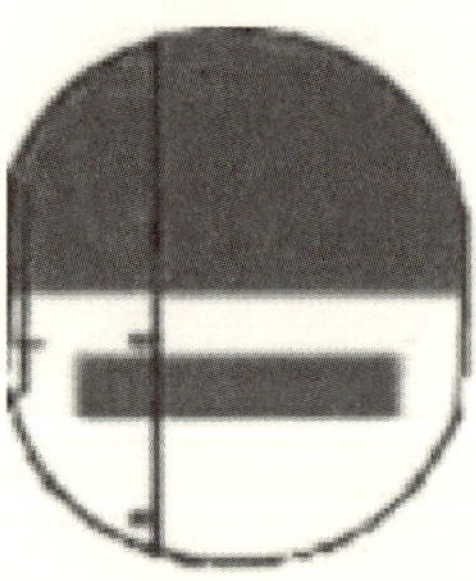

461st Tail Marking

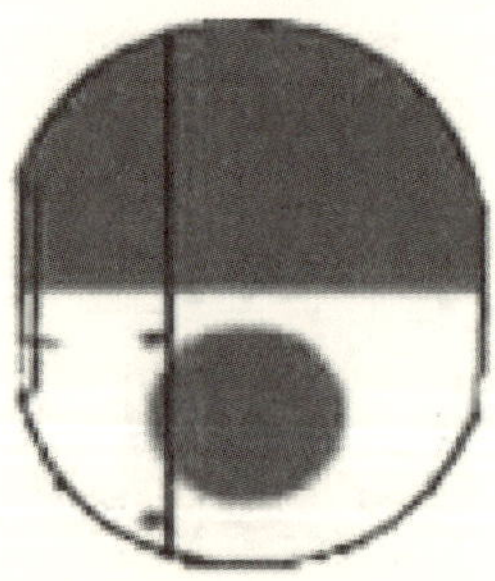

451st Tail Marking

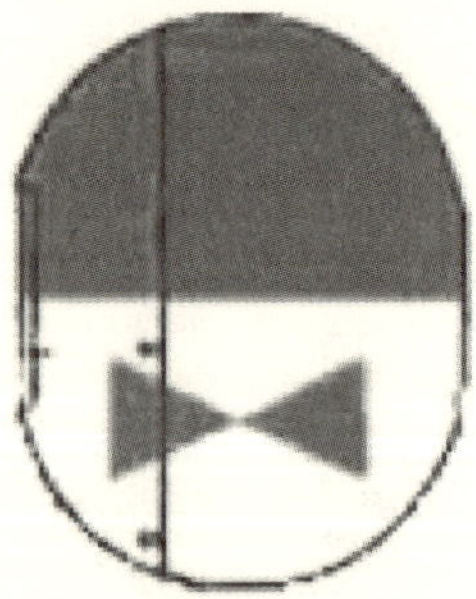

484th Tail Marking

The 49th Bomb Wing Groups flew B-24 Liberators with the tail markings shown above. The markings were red and white.

January, 1945, According to narrative history reports of the 461st Group, the month of January 1945 was one of difficult weather, too many crews, not enough airplanes or strategic targets within range of the 15th Air Force.

Most of the missions, were to targets with intense cloud coverage. Making it difficult to carry out a successful hit by the visual method.

The mission to Linz Main Marshalling Yard, Austria was recorded as the "roughest" mission of the month. The aircraft carried 100 pound general purpose bombs, the clouds were thin, but the flak was extremely intense, accurate and heavy. Twenty one of the twenty five planes were hit. Two exploded on the bomb run before they could roll out of the formation. Causing the results to be scattered.

A very special message was sent to the 15th AAF from Commanding General H. H. Arnold from Washington D. C.

"Gratifying as the results of the past year's labors have been we must not think that our job is done. The toughness of the enemy's fiber may be judged by the beating he has stood up under thus far. He will take considerably more. Our resolve for the year to come must be to hit him harder and harder, faster and faster, until his strength is finally broken."

HQ Fifteenth Air Force

From: CG, 49th Bomb Wing

To: CO, 461st, 451st, 484th Groups

The following message received from Twining cites XVIUAF A234 personal command message following a message received from Eaker.

"After reviewing the operational summaries of your strategic air force for 1944. I believe there is no organization fighting the enemy which has greater cause for pride in its accomplishments.

Your leaders have been aggressive, your combat crews have been courageous, your maintenance and support personnel have been industrious."

"Please extend to every member of your command my congratulations and great pride in their 1944 accomplishment and say that I wish ardently for each one of them continued success and safety in the New Year."

February brought about operation CLARION, launching air raids all over Germany to destroy transportation targets such as railroad marshalling yards and bridges.

Lee stated, "Maxium effort is required from all three of my groups. All three went on a mission to bomb Penzing Marshalling yards in SW section of Vienna.

Today, I attended the dedication of the 484th's new chapel that they built themselves. It was named Miles Memorial Chapel after Lt. William B. Miles who was killed in action over Elba. Chaplain Richards of the 15th Air Force delivered the address, I made a short talk after the ceremony. While there, I pinned some Silver Stars, Distinguished Flying Crosses and Bronze Stars, on the boys of the 484th Group."

"All my groups went on missions today to bomb the Airdrome at Newbigerg, Germany. All reported good results."

February 17, 1945, "I learned today that Colonel Tomhave, a great commander in the 55th Wing was shot down yesterday on a mission. I have known him for some time and think he was a good man. I hated to hear about it. I may have to let Col. A. J. Bird go to replace him."

"Today, I went up to call on Major Dooley who was hit in the face with flak the other day on a mission. He was resting OK, but will probably lose sight in one eye and the other may be impaired some. I also went to see Lt. Col. Johnston the Executive Officer of the 484th. He is in the hospital with a bad heart. He is going to be sent home."

"I had my group commanders: Col Lawhon, Col. Stefonowicz and Lt. Col. Busch in for a conference.

All my groups went on missions today and bombed Vienna South Station, but only 14 ships got to it due to weather."

February 20, 1945, " I made a short talk on the 49th Bomb Wing radio program at 10:30. The program was a very good one."

Brigadier General William L. Lee on the 49th Bomb Wing radio talk show in Italy, 1945

Lee wrote, "On February 20th, Lt. General Spratz and Jimmy Doolittle came in from England today and spent the night at Major Twining's headquarters in Bari. They will be up to Torretta Airfield in the morning.'

"I left my headquarters and drove to Torretta Airfield Lt. General Spaatz, Major General Twining, Brigadier General Born and General Curtiss landed about 10:45. I had Col. Lawhon, Col. Stefonowicz and Lt. Col. Bush, present to talk with the Generals..."

"All three of my groups went on missions today. They attacked marshalling yards in Southern Germany.."

February 28, 1945, Lee wrote, "The 461st, 451st and 484th all went on missions today. They hit marshalling yards in Ora, Bressanone and Balzone. They all did damage to their targets"

The Fifteenth Air Force continues an all out attack on the German fuel industry and communication points. Reports from headquarters state that the attacks provided immeasurable help for the ground forces.

During the month of March the 49th Wing's strategic missions continued to oil refineries, marshalling yards and other targets.

Lee wrote in his diary, "March 2, 1945 "I took off in my B-25 with Colonel Todd as copilot and Maj. John Mitchell and six other passengers for Rome. The left engine started acting up so I returned home."

Lt. Col. Sees from the 55th Wing was in Foggia and volunteered to fly me to Rome. It took us a little over an hour to get there, as we had some bad weather. My staff car was in Rome."

Brigadier General Lecel Lee with his Lancia car in Italy

March 13, 1945, Lee's diary entry stated, "All my groups went on a mission today to destroy Regensburg Main Marshalling Yards. The wing was led by Major Mixon of the 461st. They had to bomb using the pathfinder method due to weather conditions."

March 14-23, 1945 entries from General Lee's diary stated, " All my groups went on a mission to bomb Bruck Marshalling Yards today. Lt. Col Dufour of the 484th led the wing."

"All three groups bombed the Munldorf Marshalling Yards today with very good results. Another report concluded that a great deal of damage was done to the targets today."

"All groups bombed Wels Marshalling Yards today with good results. Despite eight tenth cloud coverage at the target they were able to find an opening for visual bombing.

My A-1, Chief Warrant Officer, Taylor completed 30 years of service today. He is a good man and a hard worker."

"All my groups went on missions today to bomb Kagran Oil Refinery. There was a lot of smoke over the target and the results were not as good as we wanted them to be."

During this mission Lt. Rosulek the lead bombalier was wounded just before bombs away. A plane piloted by 2nd Lt. William R. Baird was lost. He was last seen as his plane lost a propeller and had an engine on fire.

Even though there were many losses. The month of March was considered as a success. Lee's diary entry stated, "I flew down to the 15th Air Force Headquarters in Bari this morning in the B-25 to attend a farewell luncheon for Lt. General Eaker. He is the Commanding General of the Mediterranean Air Force and is returning to the States tomorrow. He will take over as Deputy Army Air Force Commander. All of my groups went on missions to bomb Budejovice Marshalling Yards in Czechoslovakia today and did a good job with 100 pound bombs."

Commendation
To: All Concerned

"The General is gratified to pass along the following message from General Arnold to General Spaatz, relayed through the general cannon: 'Noted in your message dated 24 March outlining the results achieved during the period from 16-22 March 1945. The results are source of great satisfaction to all. Such smashing aerial attacks coordinated with relentless ground pressure can only result in complete victory. Please pass to all echelons in the 15th Air Force, My Personal commendation'."

Photo Courtesy of USAF

March 25-30, 1945 Lee wrote "Today all my groups went on missions to bomb Prague Kbely Airdrome with frag bombs. The 451st did a very good job. The others had good results." A great pattern was hit across the Airdrome area and continued into the hangers destroying and damaging several enemy planes.

Today, "All my groups went on missions to bomb Strazhof West Marshalling yards. The Wing was led by Col. Stefonowicz. Good results were attained by all groups."

On the 29th "All my groups stood down today on account of the weather. Colonel Dick Warner, A-1 of the 15th Air Force was in today for a visit. I worked in my office all day."

Lee wrote, " Four Pathfinder planes from each of my groups went on a mission today to bomb Vienna North Station and Graz Depot. The 451st hit Klagenfort, the 461st and 484th hit Graz. Weather turned out good and the crews decided not to try to make it to Vienna due to the lack of fighters for protection.

They hit alternate targets using visual methods and the results were excellent."

April started with difficult cloud coverage, causing several missions to be aborted. General Lee wrote on April 2, 1945, "All my groups went on a mission to Saint Polten and did a good job."

Commendation

From: 15th AF, A311

To: Commanding Officer 461st Bomb Group

"Bomb strike photographs of your attack on Saint Polten Marshalling Yards show excellent bombing patterns. Post raid reconnaissance shows terrific destruction created. Please convey to the Groups my commendation for a job well done."

Colonel Todd

Lee's diary entry for today stated, "The President of the United States, Franklin D. Roosevelt, died at Warm Springs, Georgia this afternoon at about 1:15. He was the 32nd President of the United Stated and was in his 4th term. Vice-President Truman took over the Office.

Following the Air Force Interdiction policy the groups were directed to bomb a small railroad bridge at St. Veit, Austria. The bombing pattern was exceptionally good."

April 13,1945, "Brigadier General Herb Thatcher and I flew up to Frankfort, Germany this morning and drove around through town. It is the worst beat up city from bombing I have ever seen outside of Cassino, Italy. I went around to call on General Eisenhower this afternoon but he wasn't in. We went down and visited Brig. General T. J. Davis the adjutant General of the Supreme Command of the America Expedition Air Forces Headquarters. This is the first time I've seen him since I left the Philippines in 1938."

April 14, 1945, " I drove to the Reins Airfield and took off this morning for my headquarters in Italy. This ended a two week trip of visiting in France and England. I was a little weary. Very few clouds in the sky today.

Only one formation was able to identify the target and drop it's bombs today. Anti aircraft fire was slight."

April 15, 1945, "Lee wrote, "We had a memorial service for the late President F. D. Roosevelt in our open air theater today.

USAF Photo

Squadron Leader Michael Miller and a couple of other boys by the name of Booth and Murray were in to see me today."

Memorial Service for President Roosevelt, April 15, 1945., Italy.

According to USAF reports, "Operation Buckland continued today in support of the 5th Army. The objective for the 49th Bomb Wing was to disorganize the enemy's day. The Red Force led by Colonel Rogers was to hit enemy troop concentrations and gun emplacements about a quarter mile from the junction of the Reno and Sette Rivers.

The pay load of two hundred fifty pound bombs were dropped on the targets, leveling the installations. The Blue Force led by Major Baker Bombed the Nervessa River Bridge. They reported some success but the weather hampered the mission somewhat. However, all in all the results of the day shows excellent bombing."

Commendation

"I send my most hearty congratulations to your entire command for your record effort of 15, April, 1945, in support of our ground forces. To your ground personnel particularly great credit is due for the near miracle of maintenance which allowed 1233 heavy bomber and 629 fighters to be dispatched on that day. Operational and Intelligence staffs who conceived, planned and directed the execution have every reason to be proud. To the combat leaders and crews for a splendid day's work; I extend my appreciation and commendation."

Signed: General Cannon

Commendation

"I want you to know how deeply I appreciate the throurghly effective support which the Air Forces under your command have already given the 5th Army Group in the current attack. Your excellent cooperation began with the planning phases of our operation and has continued in a manner which is most gratifying to all personnel in the 5th and 8th Armies, which you are supporting."

Signed General Clark

Commendation

Order of Day Number 2: "The advances of our ground forces have brought to a close the strategic air war waged by the United States Strategic Air Forces and the Royal Air Forces Bomber Command. It has been won with a decisiveness becoming increasingly evident as our armies overrun Germany. From now onward our strategic air forces must operate with our tactical air forces in close cooperation with our armies. All units of the United States Strategic Air Forces are commended for their part in winning the strategic air war and are enjoined to continue with undiminished effort and precision, the final tactical phase of air action to secure the ultimate objective-complete defeat of Germany."

Signed: General Spaatz

April 16, 1945, Brigadier General Lee's diary entry for today stated, "I presented Capt. McLaughlin, my radar Officer the Air Medal this morning. Colonel Robert F. Warden, reported in for duty today. He is a graduate of Texas A&M College class of 1933, and was stationed at Kelly Field when I was there. Colonel Brooks Lawhon reported in from the 461st Group today and will take over A-3 with Colonel Ray Todd. The bad weather hindered missions today."

April 17, 1945, "I worked in my office most all day. I flew a B-25 for an hour and five minutes this afternoon. My three groups went on a mission to bomb troop concentrations in the Bologna area today and did a good job!"

A report from group headquarters stated, "Today brought an end to Operation Buckland, where the 15th Air Force had supported the armies on the ground. Photos show bombing was good and the patterns were well concentrated against the troop concentration near Bologna, Italy."

April 18-19, 1945, Lee wrote, "My three groups stood down today. I flew the B-25 for an hour this afternoon. I spoke with Colonel A. J. Bird Jr. Commanding Officer of the 464th Group today. Colonel Ray Todd my A-3 officer transferred to the 45th Group today, Colonel B. Lawhon took over as his replacement."

"All my groups went on a mission today to bomb Avisio Railroad Viaduct and had good results." On the lighter side, "I got a very nice 8 X 10 in. picture of General Eisenhower, an old friend of mine who is Supreme Commander of the Allied Expeditionary Forces in Europe."

On the 20th, Lee stated, "All three of my groups bombed Lusia Bridge today with good results. All groups did the most accurate bombing they have ever done. Lt. Col. Busch the Commanding Officer of the 484th was in to see me today at noon."

According to a report from the 484th Group, "Major Phillips led a very successful mission against the Lusia Bridge, one of the few still standing across the Adigo River. It was being used by retreating Germans. Direct hits were observed on the abutments, approaches and the bridge itself."

April 21, 1945, "I flew the Cessana for an hour with Lt. Col. John Mitchell as passenger. I had a gathering of all ex-students from Texas A & M College in the group of the 49th Bomb Wing. We had 33 people present and all enjoyed being together."

April 22, 1945, Lee's diary entry for today stated, "None of my groups went on missions today due to bad weather over the targets. Generals: Acheson, Upthegrove, Lawrence, Rush, Bill Hall, Billy Nelson and I attended the Army Boxing matches between the 15th Air Force and the Persian Gulf Command. The boys from the 15th Air Force won all the fights except one."

On the 23rd, "I flew back from Bari in the B-25 this morning. My three groups went on a mission to bomb Badia Railroad Bridge and had really good results. The Wing average was ninety one percent within a 1000 ft. radius. I flew Col. Warden back to the 461st Group.

The group reported a good account of the bombing even though they encountered heavy antiaircraft fire which damaged nine planes. They claimed twelve direct hits."

Commendation

"The results of the bombing of the Badia Railroad Bridge on 23 April 1945 are a source of great pride and satisfaction to me. The accuracy and precision displayed on this mission are indicative of the standards desired by every commander engaged in the final defeat of the enemy. My commendation to you for a job well done."

Signed: Brigadier General Lee

April 24, 1945, "Today the groups led by Major Baker bombed Robereto Marshalling Yards and gun emplacements. Thirty seven planes effectively hit the marshalling yards with their bombs. One aircraft was hit by flak.

An article from the MAAF HQ stated, "In the all out effort against enemy transportation, the 15th Air Force groups have destroyed over 4,000 vehicles in Po Valley."

Commendation

"I congratulate you and your personnel on the fine results obtained in the attack against the Robereto Marshalling Yard on 24, April, 1945. The determination and devotion to duty displayed on this attack are to be commended."

Signed: Brigadier General Lee

Commendation

"The successful bombing of yesterday, 26 April 1945, by all Groups of this command was a fine display of leadership and aggressiveness. Your action in reaching targets through adverse weather that turned back many other units is deserving of commendation."

Signed: Brig. General Lee

Commendation

Attacks against the enemy's vital communications center at Lienz yesterday show excellent results by the 49th Wing. Bombing results of the 451st and 461st Groups were particularly outstanding."

Signed Brigadier General Lee

May 6, 1945, Lee wrote, " I had a meeting of all my group commanders, their deputies, executive officer and adjutants this afternoon to discuss the problems of getting our administrative records in shape."

May 7, 1945, " The German Armies surrendered to the Allies unconditionally today and thus ended the WAR with Germany after about three and one half years of fighting. We now have to whip Japan, which in my opinion will take another year."

May 8, 1945, All 15th Air Force Aircraft Stood Down On V-E Day And Offensive Operations Ended.

General Lee wrote, "I made a short talk at the Victory Day Services in the open air theater this morning. I had my group commanders in for a meeting later this morning. I flew down to the 484th Group and presented forty medals to personnel of that Group. Made a talk on the Foggia radio at 6:30 p.m. about our victory over the Germans."

A recent report from the USAF National Museum stated, " The Allies overran Germany from the west during April 1945, as Russia forces advanced from the east. Only a few strategic targets remained for attack from the air and these were rapidly destroyed.

The last missions against industrial targets, railroads, bridges, and airfields took place on April 26, 1945. The air offensive was of tremendous value in bringing about victory in Europe."

The war ended on *May 8 at 11:01.* Celebrations started around the world as Germany signed the unconditional surrender.

May 24-25, 1945, General Lee's diary entries stated, "Capt. Cooper flew to Bari in the B-25, picked me up and flew back to my headquarters. Brigadier General Bob Acheson and Lt. Col Whitfield flew up for a party. Among those attending was Kathleen Graft Lundy, an old friend of mine from my home town of Weatherford, Texas. We all had a good time."

From J*une, 1945 until August 2, 1945,* General Lee remained with the 49th Bomb Wing while the groups were sent back to the United States.

August 3, 1945, General William Lecel Lee became Commanding Officer of the 15th Air Force in Bari, Italy. The need for Strategic Operations had ended and the wings were being sent home or reassigned.

August 5, 1945, 15th Air Force Headquarters, 31st Fighter Group with 307th, 308th, 309th Fighter Squadrons began the movement from Triolo Airfield, Italy to the United States.

August 10, 1945, General Lee wrote, "Today Japan offered to accept the terms of unconditional surrender, provided the Allies would let the Japanese Emperor remain in office."

"I flew down to Bari and went into my office at the 15th Air Force HQ. I went to the Post Exchange and bought three Rolex Auster watches, one for Col. John Mitchell, one for Major Bill Booth and one for my son, Bill. I paid $40.00 each for them. They are fine watches."

August 11, 1945, "Lt. Col. John Mitchell and I went by the America Vice-Consul to see about getting Tony Olivon to the States. It doesn't look very likely."

> *August 14, 1945*, Lee stated, "*The Japs accepted the peace terms of the United States, Great Britian, Russia and China today and thus ended the greatest war in the history of the world,"*

August 15, 1945, the diary entry for today stated, "I worked around my office at the 15th Air Force Headquarters this morning. *This morning we were officially notified that the war with Japan is over."*

August 16-17, 1945, General Lee wrote, " I drove from Bari, Italy to headquarters of the 49th Wing, eight miles south of Foggia this morning in the staff car.

Today was a holiday celebrating the ending of the war with Japan. The weather is a little hazy in Po Valley today."

Brigadier General Lecel Lee's work was done for the 15th Air Force in Bari, Italy. He remained in Italy to take on the position as Deputy Director of the Air Force Sub-Commission of the Allied Commission in Rome.

The commission was set up to oversee the domestic and military business of Italy. The Commission was headed by Air Vice Marshal Ian Brodie.

General Lee stated, "I took the job with somewhat an ulterior motive. I thought when I got the job.

I could get Mrs. Lee to come to Italy. She'd never been to Europe, but had been all over the Far East. I wanted to get her and my son Bill over here to Europe."

Lee continued, "When I got the job, I got several of my staff officers from the old wing headquarters to go with me. I got Frank Marek, Bob Warden, Hack Mixson, Tex Underwood, and Jim Graves assigned to work with me and the Royal Air Force boys."

The job didn't turn out to be what General Lee had envisioned. In his oral history interviews he stated, "I found out after I got the job, it would actually interfere with my plans to get my family over to Europe."

Major Scott, Lt. Colonel Grogan, Lt. Colonel Hawes, Major General Nathan F. Twining, Brigadier General Lecel Lee at 49th Headquarters, Foggia, Italy 1944

Major General Twining, Brigadier General Lecel Lee

Chapter 8

FIGHT IN ROME, ITALY

General William Lecel Lee recounts the beginning of BIG trouble for him in Rome. “Well, when Truman was running the haberdasher shop in Kansas City, John Maragon was a shoeshine boy in the hotel there; and he got acquainted with Harry Truman. Now, John was a Greek-born American who was sent to Greece as a special assistant the Ambassador Henry F. Grady, who was then the Ambassador to Greece.”

It is believed that Harry Vaughn sent him over there as special assistant to Ambassador Grady for the Greek elections. “So he came through Rome. He came in on a Military Air Transport airplane in early 1946.”

“He had to lay over two or three days in Rome before we had a MATS plane going into Athens. Now this is where the trouble started, “Colonel Lew Gravis was in command of the MATS operation in Rome, and he lived in the Hassler Hotel. Major General Bob Webster, commander of the MATS operation in Europe had headquarters in Paris, France. Well, Bob Webster flew in and called me at my apartment that night to tell me he was in town. I invited him to come to the house and have dinner with me.”

Lee had a penthouse apartment that had been requisitioned from Prince Massimo. He also had a staff car to drive down and pick him up at the Hassler Hotel. Lee drove to a palazzo. Lee tells his account of this slapping during an interview after he retired.

On *February 15, 1946,* "I was backing into this place to park so I could pick up Major General Webster. In the process of backing up, I backed right in front of a bus. The bus was stopped. It was a GI bus, an American bus. The guy driving was Italian. He started blowing his horn. So, I got out and asked a couple of GI's if they could speak Italian, one said he could. I asked him to find out what that fool was blowing his horn at me for?

The driver said the battery on the bus was low and if I parked there, I would be in the way if they needed to push start it. I explained to him that I was just going to be a few minutes. I'm going inside to get General Webster, and I'll be right back.

Well, by now I was kind of irritated by all this confusion and the guy blowing his horn at me. When I turned around to leave, there was a man standing right under my nose. I nearly stepped on him. I asked him what in the hell he was doing here? The man replied, 'I was doing what I pleased; I can go where I want to go.' Then I told him that may be true; if you don't go sticking your nose in other people's business. He said, 'I can stick my nose in anybody's business I want too.'

Now that was the wrong thing to tell me, cause I'm kind of hotheaded anyhow, so I slapped him twice before he could get away from me. I didn't hit him; I slapped him and he ran. I didn't run after him. I wasn't intending to mutilate him or anything.

I went into the hotel and found Colonel Gravis, Bob Webster and Dick Mazzarini and his wife having a martini. So, I went in and ordered one too.

I sat there, telling Lew Gravis that I had just slapped a guy out there in front of his hotel. I wanted him to know in case he got some complaints.

Colonel Gravis wanted to know what the man had done. I told him that the man was sticking his nose in my business, and explained what had happened. As far as I was concerned it was over.

We left to go to my place. After a while we returned to the hotel to get Col. Garvis' B-4 bag. I waited in front of the hotel. When Garvis returned he threw his bag into the back seat and got in the car. He looked at me and said, "Say, you know who that guy was you slapped out here a while ago?" I said, "I have no idea."

"Well," he said, "he's President Truman's special representative to the Greek elections" I asked Garvis, "What in the hell is he doing here in Rome? Bob said, "He's on his way to Athens, and he is waiting for a flight out." Now, I wanted to go talk to him and straighten things out. Bob said, "leave him alone, because, he's been in there trying to call Washington. So just leave him alone."

April 8, 1946, Lee recorded in his diary, "I flew to Foggia in a B-25 this morning. Today I was informed that the commanding General of the Mediterranean Theater of Operations proposes to reprimand me under the provisions of Article of War 104 for ruffling up a Mr. John Maragon in front of the Hassler Hotel on February 15th. I accepted punishment under 104 H.W.. I did not demand trial by General Court Martial."

April 26, 1946, "I attended a conference in the Air Vice Mashall's office with Commander Field and Squadron Leader Ward.

Today, I was notified by the G-1 of the Mediterranean Theater Operations that I would be reduced to the grade of Colonel on the 30th of April and that I would probably be sent back to the United States soon."

April 27, 1946, Lee states in his diary, "I talked to Brigadier Lush (British) the executive commissioner of the Allied Commission about me being ordered home right soon. He called General Hamblin and was informed that I could stay on my present assignment for awhile longer, but I heard later that I would probably have to leave with in the next few days."

April 29, 1946, "Today I heard that I may be here until the fifteenth of May."

April 30,1946, "I heard early this morning that I was not leaving Italy until May 15th. Then about noon I got a message from Cosenta that I must leave immediately, as soon as I can get priority on the ATC back to the States. I spent most of the day packing."

May 1, 1946, Lee's diary states, "Col. Miller G-1 of Mediterranean Theater Operations USA called me early this morning and told me that I was reduced to my permanent rank of Captain effective as of yesterday. I would receive and appointment as temporary Colonel from Dec. 1945. So I am no longer a general officer and probably will never be."

"Lt. Colonel Cross of the French Air Force and an Admiral of the French Navy was in my office this morning and presented me with the Legion of Honor, Chevalier Grade and the Croix de Guerre metals of honor."

"That fight cost me. I got reprimanded by Lieutenant General John C.H. Lee commander of the Allied Forces down at Caserta, Italy."

Lee said, "I was a brigadier general and probably might have wound up as a two or three star general, if I hadn't gotten mixed up in that fight."

John Maragon went on to Athens, but didn't stay long because he got caught up in a money scandal and the Ambassador sent him back to the United States. He was found guilty of lying to senators and was convicted of perjury and spent time in jail.

John Maragon

After I was sent home, I went to the Pentagon to see people in the Air Force so I could ask them what had happened and what all that business was about and they told me. Harry Vaughan had ordered Major General John Lee to reprimand me for slapping John Maragon."

"John C.H. Lee retired as lieutenant general of the Army. He wrote me a personal letter explaining why he had to reprimand me. He said, 'I was ordered from the Pentagon in Washington.' Lee stated, "I doubt President Truman ever knew one thing about it."

"I was never asked to apologize or contacted in anyway by Harry Vaughan. It appeared that he had gone to the Pentagon, and applied pressure. Perhaps in Harry Truman's name, without Truman ever knowing about it. But I knew, Harry Vaughan did not like me. I left Italy on May 15, 1946."

Back Home

In Weatherford, Texas

June 3, 1946, Lee wrote, "I stayed around Emily's house on the farm all morning. Bert, Bill and I drove into town this afternoon. We stopped by to see Ruby Dunn and Mrs. Concanges for a while. Bert, Bill and I drove Mrs. Concanges to Mineral Wells, Texas, to see Ike Concanges. Bert bought some clothes in the dress shop and we visited Ike in the Baker Hotel, had dinner and drove back to Weatherford.

I got a call asking if I wanted the Military Attache job in Manila, Philippines. I decided against it and sent a telegram saying I had rather not have it."

"The next day we drove to Cresson to see A.W. Slocum and his wife. Jeff Slocum and wife stopped by and we visited with them for a while. We returned to Weatherford and drove down to see Mr. Bill Tanner and his wife.

June 6, 1946, "We went to C. T. Talpos' house. They have a beautiful home. Later Ruby and Jack Dunn, my sister Emily and her husband met us for dinner."

Colonel Lee enjoyed a few days of rest and relaxation with family and friends, before being assigned as commander of the Smokey Hill Air Force Base in Salina, Kansas. He served there as the head of the pilot navigation training from August 12, 1946, to September 15, 1947.

Colonel Lee was now assigned to the Strategic Air Command in the United States. He said, "I went to the Pentagon when I got back to the States and checked in at The Air Force Headquarters. I talked to General Ira C, Eaker, and General Carl Spaatz. They told me I was being assigned to SAC headquarters at Bolling Field,"

Lee wrote, " Later, I talked to Major General St. Claire Street, they decided to assign me to the 15th Air Force which headquartered in Colorado Springs at that time."

"Then, they changed their minds. I was sent to Salina, Kansas. The commanding Officer there was Moe Preston, he was being reassigned. I was being sent there as his replacement."

Colonel Lee took command of the 97th Bomb Group. Lee wrote, "We were a Strategic Air Command unit that flew B-29 aircraft. We were very proud of the outfit, and we worked with enthusiasm."

Our training crews all had certain targets. They took great pride in their work, they were competitive."

January 7,1947, Lee wrote, "Major Patridge and Brig. General Freddie Smith landed at about 10:30 in a B-29. They came in from Fort Worth, Texas to meet with Major General Born, and me to discuss personnel on the station."

January 24, 1947, Lee stated, "We attended a conference in Colorado Springs, Colorado, at the 15th Air Force Headquarters all day."

February 1947, Lee wrote, "Plans were being made to send B-29 bombers to other bases. I had a phone call today from Major General Born in Washington D.C. giving me some additional information about the flight of B-29 aircraft going to Montivedeo Uraguay."

March 28, 1947, " I received a call from Major General Born informing me that Smoky Hill Army Air Field was going to be declared surplus on 31 May." Lee's diary entry for the next day stated, "I received a telegram from Army Air Forces advising us that Smoky Hill Army Air Field would be closed on or before 30th of June."

April 3, 1947, "I received a phone call from Drew Pearson about the run in I had with John Maragon in Italy. The Kansas City star called me for an interview about the same incident. The Salina Journal called for a statement about the Drew Pearson article. The Washington Post called me tonight to check up on the story about my run in with Maragon."

It just never stops. The newspapers are jumping on this story. What kind of man was he?

Little did General Lee know that the man he had slapped was well known in Washington as a "Polo Boy". Robert G. Nixon stated in an oral history interview, "If you typed con men for a movie, you couldn't do better than John Maragon.

This character, named John Maragon, was a man with a Greek- America background, who was a sort of dog robber, which was a term familiar in the First World War for a guy who did the dirty work for the captain of the company who couldn't do it himself, because of his position. Maragon was a friend, primarily, of Harry Vaughn's."

Nixon said, "I later learned that he wasn't going in to see the President, he was going to see Harry Vaughn. Although later, through Vaughn's influence, Truman sent Maragon over to Greece on some mission, and he got in trouble in Italy on his way to Greece.

This was embarrassing to the President. Maragon was throwing his weight around too much, Personal Diplomatic Representative to the President of the United States--he called himself. That's how far this idiot carried it. He embarrassed everybody in the America State Department circle in Athens. Then later having diplomatic immunity, brought back a large amount of expensive perfume into the country with no tariffs paid."

Lee continued his work with the Air Force. He was faithful and dedicated. At one point he told a reporter, "A brigadier general has no business slapping anybody." Jerry Lee took his punishment philosophically, he said, "Well if I can't be a tree, then I'll be a darn good bush."

June 1947, Lee stated, "I received the Italian War Cross for Valor and the Italian War Cross for Merit along with the Citations in the mail today."

July 5,1947, A little rest and relaxation for the Lee's included a trip home: "Bert and I drove to Weatherford, Texas this morning. We stopped by the 1st National Bank to see Fred Measures and other friends. We went by to see Ruby Dunn and Mr. and Mrs. Concanges for a little while."

August 4, 1947, Lee wrote, "Major General Charles Born called me from Langley Field, Va. to tell me that Army Air Forces had made me available for transfer to the Tactical Air Command, and that he wanted me to go to Turner Field, Albany Georgia. So I guess, I'm on my way again."

August 25, 1947, Colonel Manzo of the 15th Air Force was in today. He told me that my orders for transfer to the Tactical Air Command were on the way and that I should get them in a few days."

September 5, 1947, Lee's diary entry stated, "We left Smoky Hill Army Air Field at 7:00 a.m. We got to my sister Emily's house in Weatherford, Texas about 5:00 p.m." The next day, "We loaded the Ford down and took Bill our son to Texas A & M where he will attend the university. We unloaded his stuff and left him there."

Colonel William L Lee and his wife Bert arrived at Turner Field, GA on September 17,1947.

Lee stated in an interview, "I, not being a "fighter" man, thought it a little odd for them to put me in a fighter outfit.

As it turned out, the base had been closed, they were reopening it and putting the 31st Fighter Wing in there. So the job was administrative as well as tactical. I was sent there to be the first Commander of the 31st Fighter Wing."

Colonel Lee had not received training in fighter aircraft like the P-51 Mustang. Lee stated, "I'm commander of this outfit. I have to know how to fly this airplane." According to his records he went out on his lunch hour, sat in a P-51 while he studied the manual and instruments.

Later in an interview Lee stated, " One day during the noon hour, I got the thing up and flew it about an hour; I think half the base was out to see me land. When I came in, fortunately I made a good landing with that fighter aircraft."

Lee ended up with twenty-five hours which certified him to fly the P-51 Mustang Fighter across country. He said, "I just had to do it, if I wanted to maintain integrity as commander of the fighter wing."

October 3, 1947, Lee wrote, "I flew a P-51 for an hour this morning locally. Made one landing. It is the first time I have ever flown a P-51, and I liked it fine.

October 6 1947, "I flew a P-51 for an hour and 15 minutes this morning.

October 14, 1947, "I flew a P-51 for one hour and 5 minutes this morning, made one landing..."

P-51 Mustang Fighter

October 20 & 21, 1947, "I flew the P-51 for one hour and 15 minutes this morning. I went to witness the Tactical Air Force and Airborne Troop Demonstration today, and talked to Lt. General Pete Quesada, Commanding General of the Tactical Air Command."

Colonel William L. Lee Commander of the 31st Fighter Wing

November 20, 1947, Colonel William L. Lee was the first commander in a list of distinguished commanders, including Air Force's first jet ace, Col. James Jabara, to head the 31st Fighter Wing.

The objective for this assignment was to train pilots on P-51 Mustangs until they reached tactical proficiency. The Tactical Air Command (TAC) was established to provide a balance between strategic air defense and tactical forces of the post WWII Air Force.

The 31st Fighter Wing motto: "RETURN WITH HONOR" is exactly what they did. The Wing was prepared for f the future. They ran missions, prepared mentally and physically. A newspaper article noted the 31st Fighter Wing's competitiveness and willingness to be the best and to always "return with honor."

Colonel Lee was not only known as the 'toughest guy in the airforce," he was also known as the "enlisted men's general." He was applauded for building relationships and maintaining high morale in his wing, no matter where he was serving.

He continued as commander at Turner Field, GA until late 1948. He spent a large amount of time traveling around the states speaking on various topics.

January 1949, Colonel Lee was informed that he had been selected to attend the National War College in Washington, D.C. In the 1972 interviews, Brigadier General Lee said, "Well, Major General Don Old called me and said, he had just received notification that I had been selected as a student for the National War College.

The War College is the highest school the military has --or had at that time. I got my orders, and we went to Washington."

January 20, 1949, Lee wrote, " Today was a holiday for us on account of the inauguration of Harry S. Truman as the 33rd President of the United States. It is estimated that about one million people saw the inauguration and parade."

June 21,1949, Colonel Lee stated in his diary, "I graduated from the National War College this morning.

Secretary of Defense Mr. Johnson gave the address and President of the United States, President Truman presented the diplomas."

Immediately following graduation from the National War College, Colonel William L. Lee attended additional courses at a Radar Bombing School.

July 18, 1949, he wrote, "I am 46 years old today and I spent the day going to classes at a radar bombing school. I was informed today that I will go back to Strategic Air Command Headquarters, when I finish the indoctrination course on the atomic bomb at Saudia Base, Albuquerque, New Mexico on the 29th of July."

July 25, 1949, Lee wrote, "I saw and talked to Lt. General Jimmy Doolittle, who is now retired, for a few minutes this morning. We spent the day at Saudia Base getting briefed on the atomic bomb. This was the most interesting day I have spent in many years. I must say, that atomic bomb is really an interesting contraption..."

August 22, 1949, Lee stated, "Lt. General LeMay commanding general of Strategic Air Command came in for a couple of hours visit this afternoon. Mr. Stell of the New York Tribune called me today from Washington about the Senate investigation committee looking into my run in with John Maragon in Italy."

The next day, "I spent most of the morning answering long distance phone calls from Associated Press, United Press, radio stations, etc.. about the Maragon incident."

August 1949, Lee was assigned to March Field, California. This assignment only lasted 6 months. On *December 23, 1949,* Lee received a call that he was being moved to Ellington Field, Houston, Texas.

"Colonel W. E. McDonald called and told me that I was being assigned to Ellington Field near Houston, Texas and that orders had been issued by Air Force Headquarters." Lee continued, "I am to report to Ellington no later than *1 February 1950.* I am mighty happy to be going there under flight training command."

Colonel Lee reported to Ellington Field where according to Global Security sources, Postwar USAF Navigator Training was being put in place. Lee's venue for training once again called him to another area.

January 15, 1950, Lee wrote, "General Henry H. Arnold, Chief of the Air Forces during World War II died at 7:45 this morning at home near Sonoma, California."

January 27, 1950, Lee wrote, "I drove up to Ellington Air Force Base near Houston, Texas this morning. I went in to see Colonel Ben Starkey, the present C. O. who is retiring next month. I will be his replacement.

Our son, Bill has been up to see his girl friend the last few days as he got exempted from several of his final exams at Texas A & M College. Bill is interviewing with a doctor at Baylor Medical School for entrance into medical school this fall."

March 1, 1950, Lee stated, "I took full command of the 3605th Navigation Training Wing at Ellington Air Force Base today on General Order #5."

April 6, 1950, "We got our first T-29 aircraft in from the factory today. It is a nice looking airplane and is equipped for navigation and radar training, and includes places for some 14 students. The first class of about 50 officers will graduate on the 10th of April."

The 48 week course provided intensive classroom and in-air instruction on air navigation.

Cadets were trained in TB-25 and T-29, "flying classrooms" to give them hands-on-experience. Navigator cadets were taught dead reckoning and celestial navigation according to a report published in the USAF journal.

Aircraft used by the Navigation Training Wing at Ellington Air Force Base, Houston, Texas *USAF Photo*

June 25, 1950, Lee wrote, "The Northern Korea Communist army invaded Southern Korea today. South Korea is sponsored by the United States Government. This invasion may lead to another world war, but I hope not."

On a positive note, "..my two sisters, Thelma and Emily and Emily's daughter Mary India drove in from Waco, Texas where they have been visiting our Aunt Ollie Green. I enjoyed their visit very much. Our son Bill will start medical school on October 2nd."

Cadets continue to be trained and prepared for combat using the newest technology.

September 7, 1950, Lee stated, " I received a call from Washington this evening to tell me that Major General Doc Strather and General Ken McNaughton in the Pentagon had authorized 5% of the Cadets, who get regular commissions in the Air Force, to be Navigators on the same basis as Flying Air Cadets. That is a good move."

Colonel Lee continued his work in training at Ellington Air Force Base in Houston, Texas. He was very engaged and participated in community events. Lee's diary entry on August 28, 1950, "Dr. Kemmerer, President of the University of Houston said, 'we are going to make you an honorary faculty member. I want you to be at the commencement exercises on August 31, 1950.'"

August 31, 1950, Colonel Lee wrote, "We drove to the University of Houston, and went to Dr. Kemmerer's office to attend the commencement program. I was awarded an Honorary Professorship on the faculty of the University of Houston, by Dr. Kemmerer, President of the University. Some 800 graduates received their degrees that evening,"

Under the command of General Lee many cadets were trained in the art of dead reckoning and celestial navigation at Ellington Air Force Base. After almost three years he received word that he was being transferred back to the Philippine Islands."

Lee stated in an oral history interview, "I was sent orders from Washington D.C. to go back to the Philippines. I was going there on special request from Colonel Otto P. Weyland. He wanted, "Jerry Lee" and no one else. So, I went to the Philippines and fell heir to the 13th Air Force in Manila."

USAF Photo

According to an article in the Saturday Evening Post, Lee's fitness reports from Turner Field, The National War College, and Ellington Field all depicted a model officer - "loyal, cheerful, hardworking, rather blunt but fair in all his dealings with his subordinates."

All that Colonel Lee needed to get his star back was a change in the White House staff. Harry Vaughn no longer had a voice with the new president.

Dwight D. Eisenhower made this statement in his memoirs. "I refuse to be influenced by nothing more important than a single factor of deportment." President Eisenhower, recognized the potential and effective training abilities of Colonel "Jerry" Lee.

Lee's unconventional mannerisms which were frowned upon by some, served him well in combat.

It has been said, that a military commander must be assessed in regard to the depth of responsibilities placed upon him. Greatness cannot be summed as mere numbers on a chart, but rather, how a man responds to the need of the hour, while looking after the men serving with him.

Colonel William Lecel Lee, was a great commander when looked upon in these terms. He was once again ready to take on a new task of "warming up" the Philippine Air Force.

Sun Dial in Manila, Philippines, the largest in the world.

Chapter 9

Return to the Philippines
&
Back to the USA
1952 - 1956

September 1952, Colonel William L. Lee returned to the Philippines as Deputy Commander of the 13th Air Force under Major General John Walker Sessums Jr.

The 13th Air Force is one of the oldest, continuously active numbered air force and engaged in numerous combat assignments. At the time Lee was transferred there, it served as a staging area for people, equipment, provided training and surveillance activities to maintain a high state of readiness for contingencies.

Lee stated in an interview, "I was in Singapore on some kind of soirée down there with the British. I got a message from the Far East Air Force, through the British Royal Air Force in Singapore, telling me that Major General Sessums was being ordered back to the United States.

The chief of staff of the Air Force, General Nathan Twining, recommended that I take command of the 13th Air Force and wanted to know what my decision is regarding the command. Well, that was not a decision I had to make, I said, I'll accept."

Major General Sessums went to the base at Wright-Patterson. He returned to the Air Material Command. Lee took over as Commander of the 13th Air Force.

To the Lee's it was almost like being back home. They were certainly among friends. The Filipinos were happy to see them return.

It was during this time, while "Jerry Lee" decided to write a letter to his friend, President Eisenhower. This time the letter requested an appointment for his son, Dr. William Lecel Lee Jr..

Lee said, "I wrote him, (President Eisenhower) a letter from the Philippines. Our son was graduating from Baylor Medical School and he had an idea that he might want to do his internship at the Presbyterian Hospital in New York. It was the teaching hospital for Columbia University.

Bill had asked me to write General Ike and ask if he could stop by the White House to say hello. I knew he was always fond of Bill, so I wrote a short letter from the Philippines. I addressed it as Colonel and Mrs. Lee.

In less than a weeks time Ike wrote me back and sent it air mail. He said, 'Unfortunately, I will be out of town. I'll be in Colorado fishing for sanity reasons. Jerry, from your letter I have gleaned some information that was startling to me. I thought you were a general officer?' Then he wrote, 'I am surprised that Bill would be graduating from medical school so soon. The time has passed quickly.'"

Lee stated, "As soon as President Eisenhower found out that I had lost my star. He started an investigation.

It was about a month later, I got reappointed for a promotion to regain my general officer rank. In December 1952, Brigadier General William Lecel Lee was formally promoted to Brigadier General.

Now stationed in Manila, Lee said, "The Filipino people that I knew in the older days and some of the boys that we taught to fly were still here.

Some of them had acquired officer rank. They were really happy to see us.

They had parties and came by to see me. I was pretty well known and respected all over the Philippines."

While stationed at Clark Air Force Base the second time in the Philippines, Lee was all over the Philippines making talks, at Rotary Clubs and attending various affairs. It was said that he was probably the most popular U.S. Military man to ever serve in the Philippines.

"The reason," Lee said, "was that I associated with the people like they were my own people. I was just one of them. The one thing I never could do--was eat their ballots. You know, the eggs that are about ready to hatch, they boil and eat them."

The second tour to Clark Base was even more enjoyable than the first. Mrs. Lee said, "We traveled to Japan frequently, went to Hong Kong, Burma, Singapore, Guam and Okinawa. That was really nice to travel with him."

Clark Field located in Manila was headquarters for the 13th Air Force. Lee wrote, "It was a staging area where troops come to when coming and going from places like Thailand, Korea and Vietnam."

During early 1956, Vice-President Nixon visited the Philippines to formally acknowledge the Philippine sovereignty over American Based in that country. However, the United States held control for nearly twenty three more years.

Lee said, "I have a lot of sympathy and compassion for the Filipino people and their problems. I believe it is also in our best interest from the standpoint of global strategy and protection of the future of our own country to maintain relationships across the globe."

Philippine Army Air Force Troops Manila, Philippines

Lee returned to the United States September of 1956. He was appointed as Commander of the 3329th Technical training Wing, Amarillo Air Force Base, Amarillo, Texas.

According to the Handbook of Texas, "a two phase system of basic training began in 1956. Amarillo Air Force Base was selected as one of the bases to carry out the technical second phase.

The missile-training department was established, and facilities were expanded to accommodate an air wing of the Strategic Air Command. The base was redesignated Amarillo Technical Training Center in 1959."

General Lee remained at Amarillo Air Force Base until his retirement in 1959. His loyal patriotic service has been recognized by many.

The following excerpt was taken from a letter, written to Price Daniel, Texas Governor, from President Dwight D. Eisenhower in reply to the Governor's earlier correspondence.

Governor Daniel had written Eisenhower to request that General Lee's retirement be pushed back allowing him to be retained in the service as long as possible.

Dear Price:

Thank you for your letter concerning the promotion and retirement of Brigadier General William L. Lee. It is always gratifying to learn that a member of the armed forces has conducted himself in such a manner as to merit the keen interest which Texans have shown him.

As you may not know, General Lee is an old friend of mine, who voluntarily, and on an "out of hours" basis, gave me flying instruction in the Philippines in 1936-39. My personal regard for him and my gratitude to him have never diminished. I have always regarded him as a fine, upstanding officer.."

With warn regard, Sincerely.
Dwight D. Eisenhower

Governor Daniel's request to extend Lee's service years was not possible. Lee retired on August 1, 1959. His loyal and patriotic service is recognized by his medals and awards.

B-24 Liberator USAF Photo

Brigadier General Lecel Lee congratulates Colonel Glantzberg

49TH AIR DIVISION

Chapter 10

Medals and Honors

An article in the Weatherford Democrat (1976) stated, "Twenty Years ago they referred to him as "the toughest guy in the Air Force" His military toughness for which he was noted brought tributes from high officials in the U.S. Air Force."

Public Domain Photo

Lee was the recipient of the ***Distinguished Flying Cross*** which is awarded to those who perform acts of heroism or extraordinary achievement while in aerial flight.

General Lee was presented this award by Major General Nathan Twining commander of the 15th Air Force. for "extraordinary achievement in aerial flight while leading a wing of B-24 Liberators on a bombing mission against the heavily defended target of Ploesti, Romania on July 15th 1944 The citation reads, "he skillfully brought the entire formation directly to the target. With the immediate objectivc partially covered by smoke and clouds, a highly successful bombing run was accomplished by coordination of visual sighting and instrument control, a highly intricate process.

Under Lee's superior leadership, the bombs from his entire formation were concentrated in the immediate target area, inflicting grave damage to vital enemy installations. Turning from the target he quickly rallied his formation and despite continued opposition brought his formation through to base with minimum losses. By his extraordinary heroism, gallantry and devotion to duty, Colonel Lee has upheld the highest traditions of the Military Service."

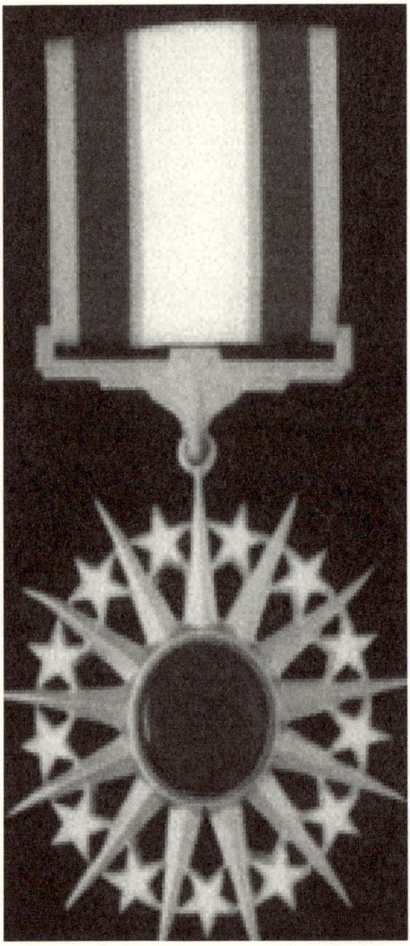

The Air Force Distinguished Service Medal is awarded to any member of the United States Air Force who is distinguished by exceptionally meritorious service to the United States Government in a duty of great responsibility. The phrase"great responsibility" means that this medal is awarded to officers who hold a rank of general officer.

United States Air Medal

Oak Leaf Clusters worn on the ribbon enlarged to show detail.

General Lee was awarded the Air Medal with two Oak Leaf Clusters. According to USAF, the Air Medal was issued for duty above and beyond and acts of heroism and bravery. Individuals who receive this Air Medal must make a discernible contribution to the operational land combat mission. During WW II, the air medal and oak leaf clusters are represented to pilots who destroyed enemy airplanes. The Air Medal/Ribbon was awarded when the fist enemy plane was destroyed. The Oak Leaf Clusters represented additional enemy aircraft destroyed. General Lee received these awards for destroying enemy aircraft during missions flown while in Italy.

An article published in the Philippine Daily Inquirer stated, "It was on May 2, 1936 that Lt. William L. Lee, a U.S. Army Air Corp. pilot, detailed with the Philippine Army Air Corp., test flew a Stearman biplane # 36-2 over Zablan Field in what is now Quezon City. Because of that particular event, May is the date set for observance of Air Force Day in the Philippines." While serving in the Philippines General Lee received the Philippine Distinguished Service Star.

According to an article in the Boeing Magazine printed 1954, "When Jerry Lee was awarded the Philippine Distinguished Service Star, it was Colonel Dwight D Eisenhower who read the citation.

Eisenhower presenting Award

The Croix de Guerre
is awarded for heroism, gallantry, as well as commendable performance during the Italian Campaign

Italian Order of Saints Maurice and Lazarus.
was presented to General Lee while serving in 15th Air Force in Italy.

Legion of Honor.
with rank of Chevalier
This is one of the highest decorations awarded in France.

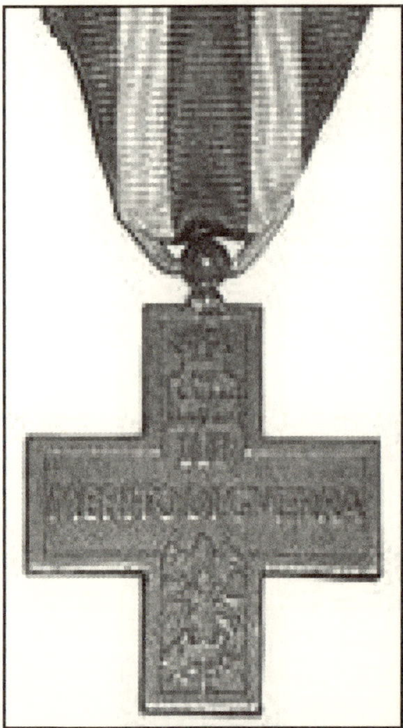

War Cross for Merit

The War Cross for Merit (Meritis di Guerra) was awarded for war merit for land sea, or air operations after at least one year's service in combat.

The War Cross a greek cross in bronze, additional meritorious service is denoted by smaller bronze crown worn on the ribbon.

War Cross for Valor

(Al Valore Militare)

The War Cross for Valor was awarded to officers for gallantry in action against the enemy.Brigadier General Lee not only received the awards shown on previous pages, but enjoyed the recognition of hard work throughout his career. This is evident by the various newspaper articles published regarding his accomplishments during his years of service.

PERSONAIRLITIES

The Aviation Monthly, August 1936

The Avation Monthly

Lt. William L. Lee

"The government or army official who was responsible for the appointment of Lieutenant William L. Lee to head the air division in the Philippine Constabulary and made him chief of the Air Corps when the constabulary was made into the Philippine Army, must have recognized a talent for organization in the massive young Army officer, who hails for the Lone Star State. Whoever it was who appointed him must have been a student of Physiognomy and must have know that those bull-dog jaws that adorn the Lee map can open up in a genial smile of pleasure or approval, and can snap shut in stern lines of reproof and authority.

The young giant who sits in the chair of the Chief of the Philippine Army Air Corps says with a ring of authority in his voice that he was born in Weatherford Texas, July 18, 1903 You'll just have to believe that, if you don't want those gimlet eyes go through you like nobody's business, even if you think he looks much younger.

Briefly, for the Chief is modest, he graduated from the Weatherford High School in 1923. Got his B.S. from Texas A & M College four years later, and graduated from the Army Air Corps Advanced Flying School specializing in Attack during 1929. He has since served in the Third Attack group and then gone back to Rudolph Field as instructor, before being sent to the Philippines. When the air division of the Constabulary was formed, he was selected to head the division and has been in the capacity ever since. You would think that a fellow like that would be satisfied with his work. He still wants to go to school! But is's not just any school he is going to. He wants to take up courses in General Staff work, for in this way he thinks he could be of more help to the Commonwealth Government.

Congratulations, Lieutenant. We like you as you are, but if you don't and you think there is still room for improvement, we can't argue that with a man from Texas. However, just so we can put in the last word, we think you've done some swell work for the Air Corps, and as an officer you're aces."

The Aviation Monthly,
August 1936

William L. Lee gave his all at what he did daily. The following quote exemplifies his character and determination.

. *"What lies before us and what lies*
behind us are small matters compared
to what lies within us. And when we
bring what is within out into the
world, miracles happen."

Henry Thoreau 1817-62

US Air Corps Photo

Colonel William Lecel Lee Presents Award For Efficiency

Mac Dill's 398th Bomb Squadron reflected justifiable pride this week and took another hitch in its collective belt as they along with the other squadrons in the 21st Bomb Group, continued intensive work on the second phase of the Operational Training Units program.

Reason for the pride was the receipt of the Group Commander's Trophy for having the highest efficiency rating of all squadrons in the 21st for the first phase of the OTU program.

When Colonel Lee inaugurated the rivalry he sent a memo that included the following, "Our present job, and it is a hard one, is the most important job this group has had and will ever have. We must train our younger and inexperienced comrades to handle their equipment so they may help us in this fight."

The Weatherford Paper 1942

Colonel Lecel Lee, of the Army Air Force, who has been supervising instructor of MacDill Field, Florida for the past few months, is now in Fort Worth for a few days on official business and visited with his sister, Mrs. Tom Tanner, on the Lee Dairy-Ranch east of town this week. Col. Lee is a graduate of Weatherford High School and has been a pilot in the Army Air Forces for sixteen years.

An Award Presentation, General Mallison, Brigadier General Lecel Lee, Lt. Colonel Fawcett

Air Attack Smashes German Army Trains

Rome, July 7 (AP).- A concentration of crowded German troop trains ready to depart for northern France was shattered Wednesday when American Liberators blasted the rail yards at Beziers in southern France. Reconnaissance photos revealed nearly 400 rail cars in the yards, at least 150 of them destroyed and many more damaged, from at least 100 direct bomb hits which dropped through heavily loaded passenger and freight trains. One explosion destroyed a train of 25 oil tankers.

The foray was led by Colonel William L. Lee, who was personally congratulated by Major General Nathan F. Twining commander of the 15th Air Force.

The Weatherford Paper

Colonel Lecel Lee Awarded Flying Cross

15th AAF in Italy- Colonel William L. Lee, Weatherford, Texas, commander of a B-24 Liberator wing has been awarded the Distinguished Flying Cross for "extraordinary achievement in aerial flight."

The presentation was made by Major Gen. Nathan F. Twining commander of the 15th Air Force, at the base in southern Italy. Colonel Lee, command pilot with 16 years flying experience in the Regular Army, led his wing on a concentrated bombing mission against a strongly defended target in Romania on July 15. "Through superior coordination and maintenance of strict flight discipline, despite heavy opposition from enemy fighters and anti-aircraft fire," the citation reads, " he skillfully brought the entire formation directly to the target.

With the immediate objective partially covered by smoke and clouds, a highly successful bombing run was accomplished by coordination of visual sighting and instrument control, a highly intricate process requiring the maximum efficiency in pilotage, under his superior leadership, the bombs from his entire formation were concentrated in the immediate target area, infliction grave damage to vital enemy installations. Turning from the target he quickly rallied his formation, and despite continued opposition brought his formation through to base with minimum losses. By his extraordinary heroism, gallantry, and devotion to duty, Colonel Lee has upheld the highest traditions of the Military Service,"

"Colonel Lee organized and brought overseas last February his B-24 Liberator Wing. Since that time, his bombers have played an outstanding part in the systematic destruction of enemy installations in Italy, Germany, Yugoslavia, France, Austria, Hungary and Greece, winning several commendations from the 15th AAF and Mediterranean Allied Air Forces, and from General Arnold, Chief of USAAF. He has organized and trained many bombardment groups both medium and heavy. At MacDill Field, Tampa Florida he organized and trained seven B-26 groups for combat between June 26, 1942 to March 21 1943. At Columbia Army Air Force Base, Columbia South Carolina, he organized the 65th Reconnaissance group. From April 25 to Oct. 29,1943 Colonel Lee was commanding officer of the 49th Bombardment Operational Training Wing. His present Liberator wing was organized in November, 1943 and moved overseas on February 4, 1944."

Lee's Photo

Adjutant reading orders awarding decorations.
Capt. Mayhowe reading. Major Besse standing behind him.
Back row: General Mallison, General Lecel Lee, Lt. Col. Fawcett

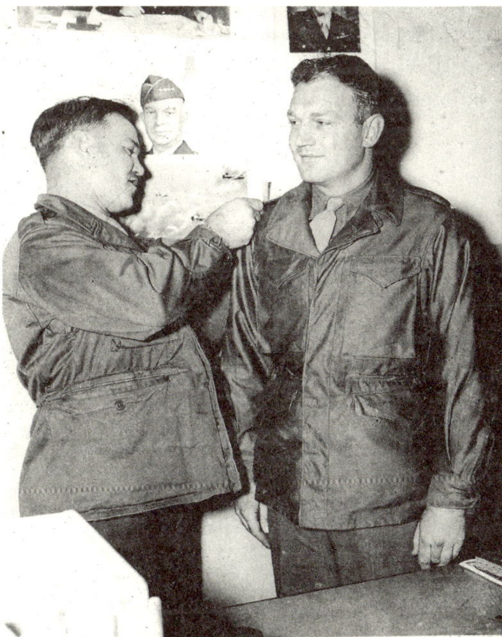

Colonel Lecel Lee pinning 2nd Lt. Bars on Lt. Pegg, who was commissioned as 2nd Lt. from 1st Sgt.

"General Lecel Lee waiting for orders to be read so he can decorate some of the men"

Bomb Groups preparing for Unit Citation in Foggia, Italy

Chapter 11

General Lee's Retirement

Lee Retires After Thirty-Four Years in Service

Printed in the Amarillo paper,

"The man who taught President Eisenhower to fly, Brigadier General William L. (Jerry) Lee retired Friday after 34 years of military service.

Lee known as the "Father of the Philippine Air Force" for his work in helping organize it. stepped down as commanding general of the Amarillo Air Force Base.

It was in the Philippines in the 1930s that Lee taught Eisenhower how to fly. The President was stationed there with the U.S. Army.

Eisenhower sent a letter to Lee to be read at his retirement ceremonies. It lauded him for his work with the Air Force and expressed the hope that retirement years would be enjoyable.

Lee was born at Weatherford, Texas. He attended Texas A & M College and graduated from the Army's Brooke Field Flying School in 1928.

Letter sent to General Lee from President Dwight D Eisenhower

THE WHITE HOUSE

WASHINGTON

July 25, 1959

Dear Bill:

I wish it were possible for me to be among those who gather at Amarillo to honor you as you retire from the United States Air Force on July thirty-first. Since this is not possible, I am delighted to have this opportunity to congratulate you upon the conclusion of an outstanding career.

Your retirement brings back to me many nostalgic memories of the service we saw together in the Philippines, and of our warm associations since. Having personal knowledge of your steadfast loyalty, energy, and good sense, I can say that this Nation has been fortunate indeed to count men of your exceptional professional qualifications among its military leaders.

Mamie and I hope your retirement years will be most enjoyable for you and Bert. I know that your concern for our armed forces will not diminish and that we may count on your unending devotion to the welfare of our country.

With warm regard,

As ever,

Dwight D Eisenhower

Brigadier General William L. Lee
Commander
Amarillo Technical Training Center, ATC
Amarillo Air Force Base, Texas

"Tribute to Brigadier General William Lecel Lee"

EXTENSION OF REMARKS

OF

HONORABLE

LYNDON B. JOHNSON of TEXAS

In the Senate Of The United States, Friday, August, 1959

Mr. JOHNSON of Texas, Mr. President, one of this country's finest soldiers has retired from the Air Force, to take up residence in Amarillo, Tex. Brigadier General William L. (Jerry) Lee has served his country with devotion and ability. He is called the enlisted man's general, and the respect his men have had for him is known even to this old Navy man.

Texas is proud to have given Jerry Lee to the Air Force, and Texas is proud to have him back again.

Mr. President, I ask unanimous consent that an article from the Amarillo Globe- Times of July 30, 1959, appear in the Appendix of the RECORD.

There being no objection, the article was ordered to be printed in the RECORD as follows:

Hard-Hitting Crusader Is Retiring Lee's Final Review Friday

(By: Jean Ely)

The invitation read: "Request the pleasure of your company at the retirement ceremony of Brigadier General William L. Lee, Friday, July 31, at 10 o'clock at the Air Force Flight Line, Amarillo Air Force Base."

General Lee, known as Jerry Lee, the enlisted man's general, will inspect his troops for the last time on Friday as they pass in review with the honor guard bearing the flags and the bands playing.

Possibly he will shed a few tears - he always does when leaving a base and old friends-but basically, he's a realist. He and Mrs. Lee will live in Amarillo as permanent residents at 4120 Paramount. He is considering a position in the public relations department of a Texas insurance company.

Retired generals don't just happen and the Nation's big companies seem cognizant of this. The 34 year record of General Lee with Uncle Sam shows great executive ability, experience, and wisdom. He received 12 military decorations including the U.S. Distinguished Service Medal, the Philippine Legion of Honor and the French Legion of Honor plus about a dozen other academic honorary degrees. When he leaves his airbase headquarters this week, he will take his gold fringed general's flag with the one star on the blue background. The incoming commander Bird. Gen. Charles H. Pottenger will put up his own general's flag.

General Lee a 200 pound muscular executive, who seems to crowd his khaki shirt, sat at his desk in the Pink Palace, as they call the base headquarters. I had just been offered a can of peanuts from which a 3-foot snake flew out when the top was removed, Gleefully, the general blew a boat horn, his crinkled face, so reminiscent of the mischievous "Skeezix" wreathed in grins. Then, sobering, he said, "Fun's over. Let's talk."

Said to be of the old military school, General Lee has been described as the "toughest guy in the Air Force." He talked of his career that had been dramatic and so legendary. He merited an article in the Saturday Evening Post when he was commander of the 13th Air Force in the Far East (one of seven such U.S. Air bases in the world). He chose a service career because he loved to fly. "I planned ahead," he said. "I figured that if I made a career in the Air Corps and retired as a major, for instance, I'd draw $375 a month pension after 30 years. In civilian life I'd have to made $100,00 to net that much retirement income at 4 percent." Consequently, General Lee refused his father's offer to enter the dairy business in Weatherford, Texas where he was born. He entered Brooks Field Flying School in 1923, after an aggressively zestful boyhood of pranks and occasional suspensions from school.

Even then he graduated from Texas A & M in the top 10 of his class. He is probably the only person who graduated with a bachelor of science in animal husbandry to achieve star rank in the U.S. Air Force.

Lee's career was a busy one, he served at 21 different military bases and was on General Douglas Mac Arthur's staff in the Philippines. Because he was the first instructor of the Filipino pilots, he was respected on the islands as the "Father of the Philippine Air Force" It was during this pedagogical period (1937) that Lee taught an inquisitive American infantry officer, Ike Eisenhower, to fly.

Like most generals who earn their star, "Jerry" Lee went to war - in the Mediterranean. His 49th Bomber Wing of four-engine B-24s reached Italy in early 1944. Lee's responsibilities were 160 heavy bombers he sent against targets all over central and southern Europe, Regensburg, Ploesti, synthetic oil plants in Czechoslovakia, and targets in Vienna, Greece, Yugoslavia and Austria.

Lee believes a superior power governs the universe. "I'd see a flier killed on his first mission, while there were many like me who survived. I only had a couple of forced landings, after 8,000 flying hours."

You can't be a fatalist in this business; it's pretty grim when a plane explodes in midair beside you through the flak."

It was as a second lieutenant in the Army Air Corps that he married Bertha Haag, a pretty San Antonio girl, who was destined to be among the most gracious and beloved of Air Force wives. The attractive Mrs. Lee is an inveterate golfer. She is a mother and grandmother. Lee's only son is Capt. William L. Lee Jr., a doctor at Andrews, AFB Washington. Capt. Lee is the father of two little girls and one boy.

The Lees have had all the glamour and protocol of the "Big Brass," Lee has subscribed to protocol in military life. Personally, he is humble and unpretentious. He believes in strict military discipline, "How can an officer command an airman of lower rank if he has been buddy-buddy with him the night before?"

In 1952 the Lees were assigned again to the Philippines where Lee was boss of the 13th Air Force. He was renowned for his sense of fun, aggressiveness and good common sense. He has never been a politician and never pussyfooted or camouflaged issues. As one observer said, "He's too honest and outspoken for his own good."

Lee still shines his own shoes, washes his own cars, darns his socks, and cleans his car's spark plugs. Since they didn't always have help, Jerry Lee scrubbed floors for his wife 14 years around the world and he helps with the dishes. "Good family morale," he stated, He's proud of his wife's skill with the needle since she has made her own clothes, including evening gowns with sequins.

As a flying instructor on many bases, Lee has seen the evolution of planes from the open cockpit, wooden propeller-DH and PTI - to today's jet that goes faster than sound.

"With the jet, we've more horsepower per pound weight and faster speed," he said, "In 1929, we used to start our own planes by pulling the copper-tipped propellers through. The planes had two wings, little wheels and a tail skid. The PT-1 had 180 horsepower. The Curtis-Wright A-3 went 110 miles per hour. Today, our planes land faster than that."

Up until 1935 Lee says the pilots wore boots, dress spurs and fancy breeches with chamois skin sewn inside the knee pads. "And some breeches were pink," said Lee, lighting a pipe. "We wore boots because most of our senior officers were ex-Calvary.

"We'd take the boots off and hang'em in the back of the planes. We flew in house shoes. When we landed, we'd put the boots back on. We also wore goggles, helmets and flying jackets since the cockpits were breezy. In the absence of radios, we'd hand notes from one cockpit to the other."

"We used to be the Army Air Corps," he continued, "but in 1947 the Air Force as we know it today was born. There were 1,800 Air Corps officers in 1938, and during the war there were 400,000 officers." Old timers percentage wise, like Lee, were a drop in the bucket.

Congress passed an automatic promotion bill to entice men into the service - I was a second lieutenant 5 years and 10 months. We had to fight for promotions until some-one died or retired.

"The promotion bill for that reason was a good thing." He stated. In speaking of his own retirement (the general is only 56) Lee said, "Officers are retired to make room for younger personnel, the service is taking in many young men and pushing out the older ones with experience, that can't be bought."

This frank, rugged militarist, with the iron gray hair that's never in place, has crusaded for many causes from the speaker's platform all over the world. He has crusaded against world overpopulation, the hazards of the installment plan, and the role of women in society.

On the eve of his retirement, he's smoldering over Amarillo's complacency about world affairs. "We've got a large growth of communism in the Unticd States. But, only one American in 100 knows what's going on in Cuba, Red China, Formosa, etc. Communism is the same as socialism," he declared bristling.

The Government is spending over 200 million out here-$40,000 on each barracks that is rehabilitated. There's a new church and new building plus the 2.5 million SAC runway." His slanting eyes that can be lost in a smile stared accusingly, "Amarillo people want to make a fast dollar; they won't rent their houses but want to sell them. We've only 500 new houses here in 900 families and there are 375 families in the mobile units," He shook his head, "When the military leaves a town because of no housing, they talk about it, which is bad publicity for the town."

The general denounced the general apathy of our business. As a Rotarian and as vice president of the chamber of commerce, he has noted that not all club members give their best to their town's interests.

Though Lee may talk gruff and spontaneous with some off trail phrases. He's a master of the written word. Last year he wrote over as thousand letters by arising at 5 o'clock in the morning. Maybe he wrote to an old Filipino cook or that the big brass in Washington. He writes to everyone who writes to him.

The general's fearless aggression has made him a debatable personality at times, but his men regard him as a square shooter. "He'll go to any extent to get his boys out of a jam, if they tell the truth," said an airman.

The Lees have attended the formalities of retirement parties, and last week Lee took his last official flight to Roswell, New Mexico.

He likes to fly because something different always occurs-like weather, engine trouble, scenery varieties and trouble you never forget. He prefers a seasoned veteran pilot to and airplane driver because a veteran usually brings a plane in during a crisis.

As General Lee walked away from his last flight, he said to himself, *"Boy, this is it!"* He didn't look back. Perhaps he was thinking about the Amarillo people who are so likeable, or that he likes this climate better than any place he's been-or maybe he was thinking of all those weeds in his new yard that is so big he may import some sheep to keep it mowed down.

Brigadier General Lecel Lee
Amarillo, Texas 1956

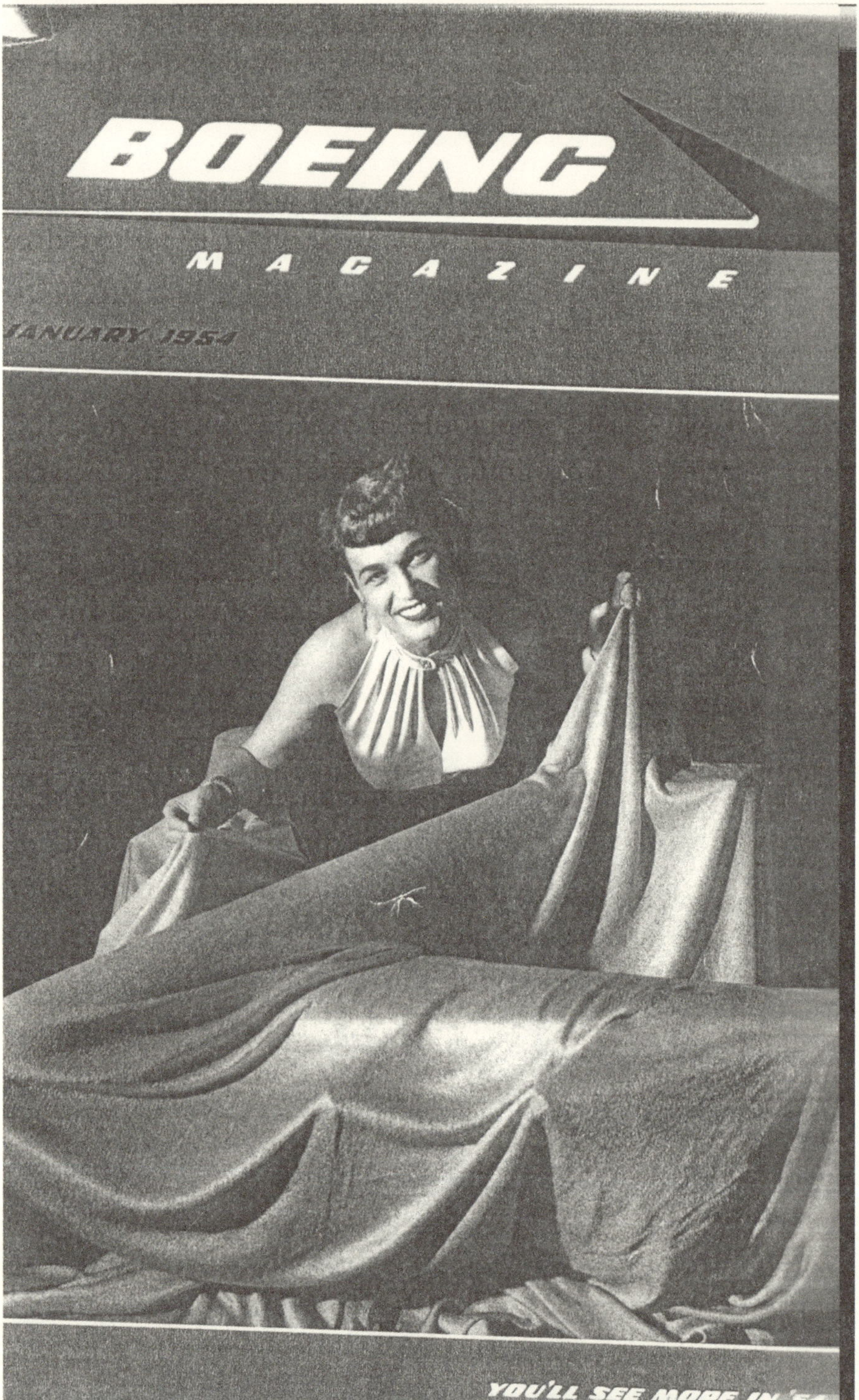

Boeing Magazine published article highlighting the life of General Lee in 1954

Chapter 12

Speeches by General Lee

It was no secret that "Jerry" Lee as his friends called him, had often spoken out against what he termed, "the domineering American female." Lee was certainly known by his boisterous attitude and opinions. But most knew, he was not nearly as hard boiled as he pretended to be. This statement was evidenced in Colonel Lee's speech delivered to the Houston Chapter of the National Secretaries Association, April 1952. The speech although a contentious topic was delivered with the traditional Lee humor and went like this:

Women

When Miss Ellis called me one night several weeks ago and asked me to talk to you this evening, I asked her what she wanted me to talk about. She said that I could talk about anything that I thought would be of interest to you.

Being unaccustomed to talking to a group of women and well realizing it would not be in order, and not appropriate to talk to you like I do a group of men, I was in a real quandary. I got my think tank to working and did a lot of serious thinking. I thought about many subjects of current interest to most men, such as corruption in our government, income taxes, the National debt, the Korean War and what's wrong with the American people.

I really racked my brain, For sure, I knew I couldn't talk to you like I do to a bunch of old stags. As you have probably heard, I usually use some strong words for punctuation, and by some, these words are classified as dirty talk.

The predicament, dilemma, plight or quandary was solved when my able and faithful secretary brought my April issue of Esquire in and called my attention to the article: "The Trouble With Women is Men." She had just finished reading it and couldn't wait to sail into my office with it, because she was in complete agreement with what the author of the article had to say. You know I have a good secretary, one who looks after my interests, one who doesn't write the letter I dictate when I am mad, but writes the rascal a nice letter the next day. Actually, she usually takes my dictation home with her and fixes it up in fancy language, telling the guy the same thing I wanted to tell him, but in nice language. I sometimes wonder why she didn't write the letter in the first place. She tells enough lies for me every day to send 50 men to hell. She reads my magazines so she can tell me which articles are of interest to me and by the same token tells me what articles are of interest to her and why she believes the author is or is not correct. We sometimes agree on the veracity, but often don't. Of course, she does the other things and odd jobs for which she is paid. Anyhow, to get back to the article: " The Trouble with Women is Men."

After reading it and rereading it and discussing it at some length with my secretary, I decided I should or would talk to you this evening about WOMEN and have dubbed my talk: "The Trouble with Women is Women" Some of you will probably not agree with all I have to say.

Some of you may want to pull out some of my remaining hair. Even thought I'm nearly a half century old, I can still run a pretty good footrace and if I'm not cornered and can make my way to the door, I don't believe I'll lose a hair, 'cause I don't believe I'll get caught.'

Getting back to the confounded article written by one young woman of whom I knew nothing. She goes on to say that American men are criticizing women for (1) being too naked and immodest (2) too intelligent and career minded (3) too undomesticated (4) too curious and too demanding of sex time (5) that women are victims of men's greed and (6) that men have exploited woman's innermost yearnings until now and men have a belly full of the results. She says-- "men have discovered that they can make money by prostituting the simple fundamental desires of women" and the lust for the dollar has undone men. She rants and raves and gives men general hell throughout the article and even goes so far as to accuse men of creating falsies. She winds up the article by saying: "Women are here to stay, they'll do anything to make a hit with a hairy ape; that women don't want to wear Pappy's britches. They don't want to rule the roost, they don't envy men and all they want to be is women--your women.

I don't believe I've ever read so much literary hog wash that provoked my contemptuous pity more. When a woman rides men like this one did, she should warble a soothing melody instead of berating us with choice billingsgate. The article certainly didn't activate my sob machine an iota. I told you that I didn't know anything about the writer of the article and I don't but I can guess a lot.

I'm inclined to believe she is one who aspires to manage the business, political and social world to suit herself.

She is probably an educated iceberg with a strong mind and devoid of sentiment as a bale of hay. I suspect that she has failed in an earnest attempt to get married-- or in a fit of desperation, married some whiskered man without sufficient moral courage to manage her.

She has probably concluded that the times are out of joint and that she has been raised up to put them right. Perhaps she is one of these women who is so skinny that her bracelets slip over her elbows and a thumb ring would make her garter. Maybe she is a perambulating tub of unwholesome lard, In any event, she appears to be morbidly sensitive to the way men are cashing in on "women's innermost yearnings"--whatever that is--she didn't say in the article. I'm not going to try to convince you that women's trouble is not men.

I don't think I would have any more luck trying to convince you than I had with my secretary. I am going to talk to ;you about women as I have observed them and tell you why I think the trouble with women is women. You probably will continue to think that some of the trouble with women is men, but, if I don't make a flop of this talk, you'll be thinking that men are just no good. You may be thinking that the problem with men is women.

Because I haven't kept up with the activities of women too close on account of doing my best to keep my head above water in doing my small bit for the Air Force, I had to research to find out what is going on in the world of women. I found out that women are working in 448 of the possible 451 known ways of making a living.

The only three, locomotive engineer, firemen and firefighter are still "men only" jobs in our country today.

I discovered that a women has risen from stenographer to vice-president of American Airlines.

Don't know how long it took her to make that climb. She is there and doing a bang up job. Several of our big department stores throughout the country have women executives. One in New York is headed by a woman who rose from the ranks. That store has more female executives than male. Over 1,000 women have commercial pilot certificates and two of these hold airline transport ratings. Parenthetically, I might add that I would not have any qualms about riding as a passenger in any transport they were piloting.

More than 400, including an ambassador and a minister, serve in our diplomatic service.

And we have a woman as assistant secretary of defense who is doing a fine job. So I found out during a recent sojourn in Washington, women serve in Congress, 236 State Legislatures to be exact. Women mayors rule Sacramento, California, Portland, OR. Red Bluff N.J.. and a dozen smaller cities some having all female administration. Nearly 200,00 women run farms.

Can a women get ahead in a 'man's world'? When a woman is asked, if women can get ahead in a man's world, she will more often than not give a bitter answer---"We don't have a chance." All we get is what men let us have-poor pay and dead end jobs. The figures quoted above about employment of women indicate that such answers as given are not entirely justified. The fact that there is something wrong with women, at least some of them, is pretty plain.

In a recent survey conducted by the Department of Labor, 646 women were asked: “ Five years from now, what position would you like to hold?” Twenty-three percent said they hoped to move up a notch.

Recently a number of women were asked how they would feel about having women supervisors instead of men. Some showed strong prejudice against their own sex. One said, Men bosses are “less afraid of able subordinates.” Another thought women are “too fussy and domineering.” Such prejudices hurt the women who hold them. They feel defeated. They can’t aspire too high jobs because their sex disqualifies them. Their prejudice kills ambition and they have no incentive to prepare for better jobs.

Is there something wrong with women? Well, I should snicker! Because some women have made great success in life and most have not made much of a success. I will not say that something is wrong with all women. The differences which show up between women in daily life are partly due to the use they make of the capacities they have. Women with an intense desire to make good- use their power fully and take advantage of every opportunity. Others drift along wit the tumbling weed although they have the capacity to do more. It is not often than not that the push or motivations is lacking, NOT brains. Women, too often are reluctant to shoulder the big responsibility that goes executive positions and better paying jobs. Too often they work with a pay check dangling before one eye and a new hat fluttering before the other while thinking about how they are going to snare that “hairy ape” they met at some shindig the night before.

Nearly half of them, 47 percent said they just wanted to hang on the same old job.

Hoping for more responsibility and more pay. Many of them said they did not hope to advance because they know that men would not let them. Discontented women are quick to point to masculine prejudice.

Men have no monopoly on prejudices because women have them too. Quite often they are carbon copies of the prejudices which women resent most in men.

By now you probably think I am a woman hater, but such isn't so. Actually, I'm very fond of women. I have many female friends, some white haired grandames, some mere girls in short dresses, some skinny, some fat, some intelligent and some not so intelligent. To me, the friendship of a noble woman is a real asset.

I don't believe there is any difference between men and women in general intelligence. Though some men, and perhaps most men seek out women of lesser intelligence in order to protect their own self-esteem, IMAGINE otherwise.

I have found that the average woman excels the average man in social intelligence. She is more tactful, less blunt, than a man. She is better at side stepping information which might show her up at a disadvantage. She will not lie more than the average man--or less-- but her lies are more subtle and difficult to detect. She needs more encouragement than men and her boss will get better results if he praises both good work and good intentions. She does not change her mind more often than men do. But it generally take her longer to make it up. When she makes up her mind it is most difficult to change. The average woman's stomach is a bit larger and digests food more rapidly, so she is more likely to become hungry between meals and need a midmorning snack.

She is more likely to be loyal to her family or to her employer and quite often had rather work for a nice boss for less money than work for a rough and uncouth boss for a fancy salary.

There are plenty opportunities for working women to make a success in careers today. Their chances of making a success are just as good as men. In order to make a success in a career, women must do some of the following: (1) Develop more than a pay check attachment for the company for which they work. (2) Forget the hairy ape while on the job-chances are he is some lug-headed dope that hopes to get married to a working girl so she can support him. (3) Learn as much about the company they work for as it is possible- what they make, how they make it, who buys it, etc. (4) Be willing and eager to accept responsibility and demand authority along with the responsibility. (5) Don't watch the clock too much-you might become one of the hands, (6) Take advantage of every opportunity for self-improvement, socially, mentally and physically.

Women don't make perfect secretaries. As a matter of fact there are no perfect people. Perfection is a relative thing, and it depends on who is judging. What may be an outstanding secretary to me may not be so good in the eyes of another man. But there are several things I like to have in a secretary. I'll list them in the order of their importance. (1) Intelligence--I'm probably an exception to most men because I want a secretary who is smarter in every way than I am. I feel like she will be able to teach me a trick or two and while so doing she will give other people who talk to her the idea that her boss is pretty smart too.

Women know darn well that a man is not going to put up with having a smarter woman than he is out talking about it. (2) I want a secretary who can detect my shortcomings and at the same time have the courage to tell me about them.

In other words keep me on the right track and out of trouble. (3) Loyalty ranks high with me and probably should come at the top of the list. Come hell or high water, I want a secretary that will stand by me through thick and thin and not behind me and see my belly gets beaten black and blue. (4) I don't want any "yes gal" working for me. If she thinks I'm wrong, I want her to tell me about it and argue her point if necessary,

I'd rather not have knock-kneed, bow-legged or snaggle toothed gals. Don't want'em too fat, too skinny, too tall or too short. Really, I want one that most men would like to grab, love up one side and down the other, just so long as she is smart enough to keep them from doing it, and if they out smart her, strong enough to poke'em in the nose.

Women

She's and angel in truth, a demon in fiction;
A woman- the greatest of all contradiction.
She's afraid of a cockroach, and scream at a mouse.
But she'll tackle a husband as big as a house.

She'll take him for better, she'll take him for worse,
She'll split his head open and then be his nurse.
And when he's well and can get out of bed,
She'll pick up a tea pot and throw at his head.

She's faithful, deceitful, keen-sighted and blind,
She's crafty, she's simple, she's cruel, she's kind.
She'll lift a man up, she'll cast a man down,
She'll crown him her king, she'll make him her clown.

You fancy she is this, but you find she is that,
For she'll play like a kitten and bite like a cat.
In the morning she will, in the evening she won't,
And you are always suspecting she does, but she don't.

Quod Erat Demonstrandum-
"Which was to be proved"

Speech delivered to the Houston Chapter, National Secretaries Association, April 15, 1952. Houston, Texas

Lee Speaks At Pelzer Rally

(The following article in regard to Colonel Lecel Lee is taken from a South Carolina paper)

It's better to buy bonds than to wear them, Colonel William Lecel Lee commanding officer of the 39th Bomb Wing at the Greenville Army Air Base, told several hundred Pelzer mill fathers of service men, at an anti-absenteeism rally Thursday night.

Honoring a number of their sons and friends who recently returned from the different fighting fronts, the Pelzer employees heard Colonel Lee's keynote speech, which pointed out that they, were the vital reserve corps needed by our armed forces to win this war.

Colonel Lee stressed that only through the loyal and determined efforts of the parents, families and friends at home can the United States hope to win this war with less blood and smaller total cost. Every bond bought and every minute producing 60 seconds of vital production helps cut the war's life line the Colonel added. He also quoted portions of a personal letter sent to him by General Dwight D. Eisenhower citing the need for home support.

The rally was held in the Pelzer Community building. Those attending had worked an earlier shift during the day so they would lose no time from their jobs.

Christmas party for kids 23 December 1944, Italy

General Lee spoke to the men in service during his tenure in Italy. He maintained a radio program aimed at helping the airmen and women serving in that area during World War II. He helped provide gifts for the children at Christmas time.

He was also well known for his Poker-Room for the pilots. The rules were in tact, His activities were noted and he received recognition for high morale in the 49th Bomb Wing in Italy.

The 461st raising the first flag over their base camp in Foggia, Italy 1944

Colonel Lee presenting awards to the 49th Bomb Wing in Italy during WWII

Chapter 13

Pictures from Lecel's Personal Album

Natives gathered around my airplane at Daja, Philippines

The man behind the little girl is Datu Paglas, Amoro Chieftan at Buluan. The woman on the right is Lt. Navarro's wife. Navarro is Constabulary Officer in charge of the soldiers stationed at Buluan

Lee, "My Lancia Auto"

An Ox Cart-taken near Zamboanga. There are not native cattle, but originally came from India.

Unit Citation in the Philippines

A group of Philippine Constabulary Officers taken at Jolo on the 12th of July, 1935

View of the Tombs in the Chinese Cemetery

A ferry crossing the river, the boats are cut out of logs from big trees.

Caraboa pulling a cart, a favorite means of transportation with native tribes, April 6, 1935

A carritella, several thousand of these in Manilla, People pay as much as ten Centavos to ride in them.

A native house (Nepa Hut) and another native participating in the Good Friday Ceremonies. Philippines

Good Friday, religious ceremonies -men are carrying crosses made out of green bamboo.

Old Spanish Cannon, It was brought to the Philippines during the Spanish time. Made in Spain, 1783.

Idea taken from China-people live in these little boats. They beg food from big boats.

Natives of the Head Hunting Tribe, Philippines

Getting a haircut in China 1935

Entrance to Nichols Field in the Philippines

A fish trap in water 18-32 feet deep. I took this from a big boat on August 22, 1936

The Pasig River - Looking to the East.

PT-2 Aircraft

Colonel Lecel Lee immediately after return from a mission in Italy.

USAF Photo published in the Saturday Evening Post March 1955
Brigadier General William L. Lee (Second from right) appeared as "Miss Hickum" (Hawaii) in an officers' "beauty contest" at Clark Field in the Philippines

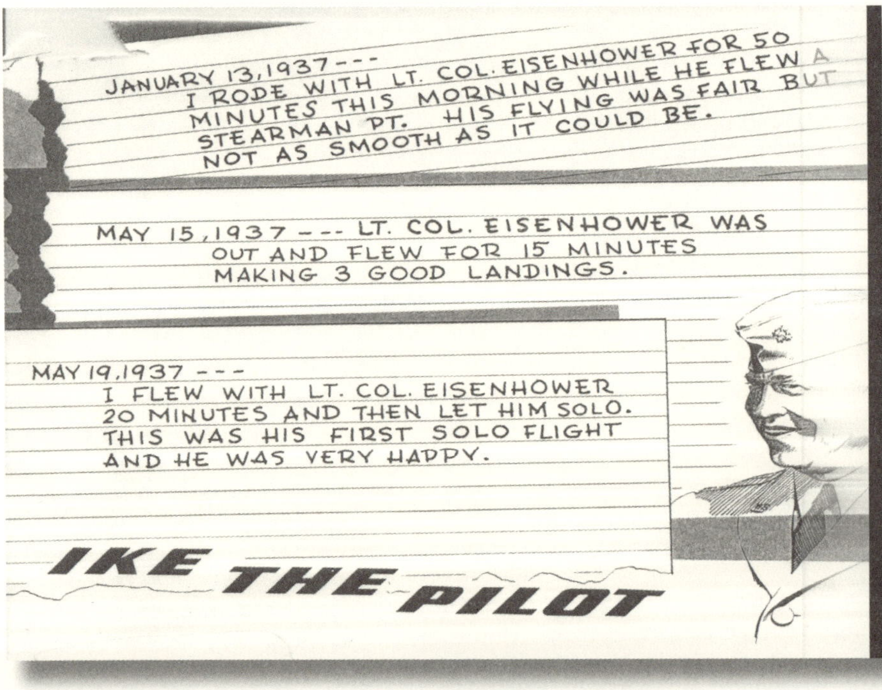

Chapter 14

General Lee's Death

The Weatherford Democrat, headlines,
May 1, 1976

Jerry Lee Remembered For His Toughness

"Twenty years ago they referred to him as "the toughest guy in the Air Force." His military toughness for which he was noted brought tributes from high officials in the U.S. Air Force, but the late Retired Brigadier General William L. (Jerry) Lee was also known and respected for his dedication to civic service as well.

Before his death on Feb. 26, 1976, Jerry Lee, 72, had earned numerous plaques and citations for work with civic groups in Amarillo.

Lee served as vice president of the Amarillo Chamber of Commerce for three years, served on the board of directors for the Amarillo United Fund for three years, served on the YMCA board, the Red Cross board, Multiple Sclerosis board and the Tri-State Fair board.

General Lee was a member of the Rotary Club, served as president of the Amarillo Humane Society and in 1968 had been appointed to the Amarillo Board of Zoning Adjustment.

During his military career he received 12 military decorations including the U.S. Distinguished Service Cross Medal, and the French Legion of Honor."

"Lee is survived by his wife who lives in Amarillo and two sisters, Mrs. Emily Tanner and Mrs. Thelma Hall both of Weatherford , TX."

Weatherford Newspaper

Brig. General (Ret) William L. "Jerry" Lee

Brigadier General (Ret) William L. "Jerry" Lee, 72 died Thursday, in an Amarillo hospital following a short illness.

General Lee was born in Weatherford in 1903, and graduated from Weatherford High School in the 1920's

Survivors include his wife, Bertha Lee of Amarillo; one son, William L. Lee, Jr. of England, three grandchildren; two sisters, Thelma Lee Hall and Emily Lee Tanner, both of Weatherford, TX., and one niece, Mary India Schoonover, also of Weatherford, TX.

General Lee graduated from Texas A&M in 1927 and was a career Air Force officer, having served 31 years.

Lee commanded the 49th Bomb Wing of the 15th Air Force during World War II.

Lee was a visitor in Weatherford several years ago for a program, held prior to the old red brick school building on South Alamo Street, being torn down."

◇◇◇◇◇◇◇◇◇◇◇◇◇◇◇◇◇◇◇◇◇◇◇◇◇◇◇◇◇◇◇◇◇◇◇

From A to Izzard - *Amarillo Globe Times*
RETIRED BRIGADIER GENERAL
William Lecel Lee

"Twenty years ago they called him the "toughest guy in the Air Force.

A couple of years ago we talked to him on the telephone. He had never left Amarillo and had turned out to be one of our town's most productive civic workers.

"I am President of the Humane Society," he told us. On another occasion he explained to us, "If you can't get me at home, try the Red Cross, I'm doing a lot of volunteer work for them."

Yesterday "The toughest guy in Air Force" was called into permanent retirement by the Man Upstairs.

Jerry Lee had held practically every major civic chore this town had to offer, for these services he received many plaques."

Among the tributes that came his way during active service, was an informal statement from Gen. Nathan F. Twining, Chief of Staff of the U.S. Air Force. It concerned a reinstatement of rank, General Lee had been cited for slapping a political representative of Harry Truman's while in Rome, not knowing who the fellow was.

"I was glad to be in a position to promote him again," said Twining. "Jerry Lee is a fine combat officer and if I had to go to war again. I'd like to have him with me."

Earlier, General Twining had reported to the Pentagon after action over Italy, where Lee had served under Twining, "I consider Lee in the first eight of 116 general officers known to me."

What America needs today is more of that toughness for which General Lee was noted. It may be that the personnel of our armed services today are not tough enough in this politically oriented world.

But give us trouble, which might well come more violently than any of us suspect, and we will be begging for a few Jerry Lees.

After all, military toughness is nothing but disciplined common sense applied promptly and firmly." At the time of General Lee's death his only son was stationed at Royal Air Force Base, Lakenheath England,

THE STORY CAN NOW BE TOLD

Retired Air Force General "Jerry" Lecel Lee of Amarillo, Texas has died at the age of 72. He received many military decorations during his faithful service to his country and from his history, comes a very interesting story that should be of interest to all of us. In 1935 Jerry Lee was a Lieutenant in the Philippines trying to make a "full grown" air force out of very raw materials. It was all a part of General Douglas Mac Arthur's idea, Jerry had his hands full. One day an American Lieutenant Colonel came to him and asked for an airplane ride. This was against military regulations but how could a Lieutenant say 'no' to a Colonel? If General MacArthur should find out they were using planes assigned to another government to play around, they would all be in trouble. However, the Colonel kept hanging around the flight line and kept asking so one day in April, Jerry finally agreed and took the Colonel for a ride.

If he thought that windy half hour in a patched up old trainer would discourage the Colonel, he was wrong. Now the Colonel insisted he wanted to learn to fly. From then on, this infantry Colonel was at old Camp Murphy's hanger and when the day's official duties were done, Jerry and the Colonel would climb into an old plane for another hour of flying instruction.

Then one day Jerry stayed on the ground and his student took off alone. Of course they were still careful not to let Gen. MacArthur find out about this unorthodox extra curricular activity.

A few months later in 1938, a newer model PT-1 was delivered to the Philippines. Lee stated many times that on this day he rode with the Colonel for 50 minutes in a Stearman PT-1. The Colonel learned to fly and became a licensed pilot. This determination to participate to the full intent, whatever he attempted regardless of the risk, is a part of the pattern of his great success.

The Colonel became the President of the United States, Dwight D. Eisenhower.

Lt. Lecel Lee

We know without a doubt that William Lecel Lee's life story is unique and uplifting in many ways. He never stopped to contemplate failure, nor did he allow circumstances to keep him from doing what he knew he needed to do. There is no better way to sum up the reflection of a great combat officer, than to print this poem found in his photo album.

Don't Quit

When things go wrong as they sometimes will,
When the road you're trudging seems all uphill,
When the funds are low and the debts are high,
And you want to smile, but have to sigh,
When care is pressing you down a bit -
Rest if you must, but don't you quit.

Life is queer with its twists and turns,
As every one of us sometimes learns,
And many a fellow turns about,
When he might have won, had he stuck it out.
Don't give up through the pace seems slow -
You may succeed with another blow.

Often the goal is nearer than
It seems to a faint and faltering man;
Often the struggler has given up
When he might have captured the victor's cup;
And he learned too late when the night came down,
How close he was to the golden crown.

Success is failure turned inside out-
The silver tint of the clouds of doubt,
And you never can tell how close you are,
It may be near when it seems afar;
So stick to the fight when you're hardest hit-
It's when things seem worst that you mustn't quit.

Author unknown -may have been written by General Lee.

"Jerry" Lee was very involved in the community of Amarillo, Texas. This clip from the Amarillo News is from his run for State Senate during 1959. Lee did not win, but continued his involvement and interest in community and state affairs.

Bibliography

1. *Oral History Interview.* (1970-72). General William L. Lee by Alan Thompson, Maclyn Burg and William Jones. Dwight D. Eisenhower Library Abilene, Kansas

2. Sherrod, Robert. *(March 26,*1955). Saturday Evening Post; *Toughest Guy In The Air Force.*

3. Hasdorfs, James C. (Dec. 1943). Exerts from "Jerry Lee"; *Founding Fathers of the Philippines.*

4. *Weatherford Democrat.* (9/30/1943).

5. *South Carolina Paper.* (1943).

6. AAF-RAF.*Public Relations Speical Handout.* (12/9/1944).

7. *Boeing Magazine.* (January 1954)

8. Lee, General William L. *Daily Diaries* (1935-1954)

9. Lee, General William L. Speech:*"Women".* (1952).

10. *Amarillo Daily News.* (1969).

11. *Weatherford Democrat.* (4-4-1993).

12. *Weatherford Democrat.* (5-16-1976)

13. *MacDill Field News.* (1956).

14. *Memorial Services, A to Izzard.*(1976)
San Antonio, TX.

15. *Amarillo Globe Times.* (March 27, 1973).

16. Ft. Worth Press - Promotion

18. *Clark AFB News.* (1954)

19. *15th Air Force*- Italy-News Article

20. Family information from Mary India Schoonover- Niece of General Lecel Lee.

21. Neighborhood information from Leon Tanner, Jack Borden and Ben Schoonover.

22. *The Melon Vine.*(1923)
Weatherford High School Annual.

23. *The Longhorn, (*1926) Texas A&M Annual.

24. *History of the 31st Fighter Wing*
US Air Force Link, www:af.mil/news/story. (accessed Oct. 2008).

25. Dickie, George. editor. *461st. (Jan-Dec 1944)*
www.461st.org/history/history.

26. Http://www.commons,wikimedia.org.

28. *484th Bomb Group.* www.484th.org/history. (accessed Dec. 2008)

29. *15th Air Force History.* www. 15th Air Force.org/ history (accessed Dec.2008).

30. *D-Day.* www. Paul.rutgers.edu/wwii/usaf/d-day

31. *Handbook of Texas Online,* s.v.”,”www. tshaonline.org/handbook/online/articles/AA/qba1. html. (accessed Decembr 8, 2008).

32. Tarolan,Ramon J.(2008). *Reflections on Air Force Day (Revelle)* Philippine Daily Inquirer: 6-23-2008.

33. *The Pilot. (*1954). www.time.com/time/magzzine/ article. (accessed Nov. 2008)

34. Brigadier General William L. Lee Biography. United States Air Force, . www.af.mil/bios (accessed Oct. 2008)

35. Lee, Lt. Lecel. (December, 1937). *Chief Philippine Army Air Corps; Gen.Santos, Col. Segundo and Lieut. Lee, Lost and Found.* Khaki and Red. Philippine Army Journal.

36. *Ploesti Raid in 1943.* USAFHS-65. Chap. 1 Declassified IAW EO12958.

37. *3605th Navigation Training Wing. Ellington AFB.* Houston Chronicle. June 23, 1948.

38. Snyder, Louis.L. (1958). *World War II.* Franklin Watts Inc.

39. Adams, Simon. (2000). *WWII Eyewitness.* DK Publishing, Inc.

40. Pietirusza, David. (1999). *Invasion of Normandy, Battles of World War II.* Lucent Book, Inc.

41. Dank, Milton. (1984). *Turning Points of WWII.* Franklin Watts, Inc.